Some RNA Viruses

Edited by Yogendra Shah
and Eltayb Abuelzein

Published in London, United Kingdom

IntechOpen

Supporting open minds since 2005

Some RNA Viruses
http://dx.doi.org/10.5772/intechopen.82983
Edited by Yogendra Shah and Eltayb Abuelzein

Contributors
Syed Lal Badshah, Khalid Khan, Nasir Ahmad, Abdul Naeem, Aqsa Farman, Nail Khammadov Ildarovich, Yaxin Wang, Meijun Liu, S.D. Audarya, Manuel Bicho, Andreia Matos, Alda Pereira da Silva, Joana Ferreira, Carolina Santos, Maria Clara Bicho, Adamu Ishaku Adamu Akyala, Chanda Siddoo-Atwal, Rafaat Hussein

First published in London, United Kingdom, 2021 by IntechOpen
IntechOpen is the global imprint of INTECHOPEN LIMITED, registered in England and Wales, registration number: 11086078, 5 Princes Gate Court, London, SW7 2QJ, United Kingdom
Printed in Croatia

British Library Cataloguing-in-Publication Data
A catalogue record for this book is available from the British Library

Additional hard and PDF copies can be obtained from orders@intechopen.com

Some RNA Viruses
Edited by Yogendra Shah and Eltayb Abuelzein
p. cm.
Print ISBN 978-1-83962-925-9
Online ISBN 978-1-83962-926-6
eBook (PDF) ISBN 978-1-83962-927-3

For EU product safety concerns:
IN TECH d.o.o., Prolaz Marije Krucifikse Kozulić 3, 51000 Rijeka, Croatia, info@intechopen.com or visit our website at intechopen.com.

Meet the editors

Dr. Yogendra Shah works as a Senior Research Microbiologist and Lecturer at the National Zoonoses and Food Hygiene Research Center and Kathmandu College of Science and Technology, Nepal. He completed his Ph.D. in Veterinary Medicine (Bacteriology) from the Graduate School of Veterinary Medicine, Hokkaido University, Japan in 2017. His research areas are mainly to better understand the molecular epidemiological features/transmission dynamics of infectious diseases and zoonotic infectious diseases in Nepal by employing molecular techniques like ELISA, polymerase chain reaction (PCR), loop-mediated isothermal amplification (LAMP) and DNA sequencing. He was awarded the 'Young Science and Technology Award' from the Nepal Academy of Science and Technology (NAST) in 2019. His research interests include infectious diseases, zoonotic infectious diseases and vector-borne diseases. He has published more than 34 research articles and 12 books in peer-reviewed journals. Publication link: https://orcid.org/0000-0002-0820-4556

Professor Eltayb Abuelzein graduated from the University of Khartoum, Sudan, with a B.V.Sc. in Veterinary Medicine. He obtained his Master's and Ph.D. degrees in virology from the University of Reading, Britain. He also received his Post-Doctoral Fellowship (FIAEA), in Immunolabelled Techniques & Arbovirology from the World Reference Laboratory, Pirbright Britain. Since 1975 he has been engaged in active research, technological developments, and teaching in several universities in Sudan and Saudi Arabia. He has acted as a Consultant to the International Atomic Energy Agency (IAEA) in Immunolabelled Techniques. He is an editor and reviewer for a number of scientific journals including the international journal "Tropical Animal Health & Production", Britain. He has published 123 peer-reviewed papers in reputable scientific journals and 60 peer-reviewed articles in conferences. He was the first to develop the ELISA technique in Foot and Mouth Disease (FMD) Virus Research. He has been quoted in the highly reputed international scientific journal "Nature". He contributed to the book: "Immunoassay in the 80's", 1981, ATP Publishing Co., London. He made the first records of several viral diseases of great international significance. Presently, he is the Chair Professor of Viral Haemorrhagic Fevers, King Abdulaziz University, Saudi Arabia.

Contents

Preface

Medically important RNA viruses with families include: Togavirus, Bunyavirus, Flavivirus, Filovirus, Arenavirus, Rhabdovirus, Paramyxovirus, Orthomyxovirus, Coronavirus, Retrovirus, Picornavirus, Reovirus, and Calicivirus. Some RNA viruses contain double-stranded RNA (Family Reoviridae) and some viruses have a life cycle that uses reverse transcriptase. Viruses contain ribonucleic acid (RNA) as their genetic material. Particularly, several human diseases are caused by RNA viruses. The main objective of this book on RNA viruses was to understand the RNA virus i.e a seroprevalence study based on molecular epidemiology such as genotyping tools by IgM/IgG, ELISA, PCR, FRNT50 and phylogenetic analysis to further confirm the RNA viruses causing diseases in tropical and subtropical countries. However, pathogenesis mechanisms causing viral diseases among the flaviviruses family have not been clearly understood and also little is known about the host responses to RNA viral infection. This book will surely help with these information gaps and also provide research information to the policymaker or planner for further diagnosis, control and prevention in future outbreaks of RNA virus diseases in tropical and subtropical countries.

Yogendra Shah
National Zonosis and Food Hygience Research Center,
Nepal

Eltayb Abuelzein
Central Veterinary Research Laboratories,
Sudan

Chapter 1

Mode of Transmission and Viral Shedding of SARS-CoV-2: Emerging New Paradigms

Adamu Ishaku Akyala

Abstract

SAR CoV-2 is an important group of animal and human pathogens that infect respiratory tract, hepatic, gastroenterological, and nervous systems of mouse, bat, bat, humans and other vertebrates. Middle East Respiratory Syndrome (MERS) and severe acute respiratory syndrome (SARS) Outbreaks in 2002–2003 have demonstrated the possibility of human to human transmission, animals to humans transmission of the emerging SARS-CoV-2. The World Health Organization (WHO) On 12 January 2020 renamed novel coronavirus infectious disease (COVID-19) to SARS-CoV-2 In late 2019, the first case of the COVID-19 was reported. A total of 87,137 confirmed cases globally, 79,968 confirmed in China and 7169 outside of China, with 2977 deaths (3.4%) had been reported by WHO in March 1, 2020. Meanwhile, several independent research groups have identified that SARS-CoV-2 belongs to β-coronavirus, with highly identical genome to bat coronavirus, pointing to bat as the natural host and by proxy has a zoonotic propensity. Angiotensin-converting enzyme 2 (ACE2) is the same receptor been used by the novel coronavirus as that of SARS-CoV and largely spreads through the respiratory tract. Currently, there are few specific antiviral strategies, but several potent candidates of antivirals and repurposed drugs are under urgent investigation. In this review, we summarized the latest research progress on the transmission mode dynamics and viral shedding in provide direction for isolation protocol. R0 estimates for SARS have been reported to range between 2 and 5, which is within the range of the mean R0 for COVID-19 found in this review. Due to similarities of both pathogen and region of exposure, this is expected. On the other hand, despite the heightened public awareness and impressively strong interventional response, the COVID-19 is already more widespread than SARS, indicating it may be more transmissible.

Keywords: coronavirus disease 2019 (COVID-19), transmission, clinical characteristics, viral shedding

1. Introduction

In late December, 2019, an epidemic of respiratory disease caused by severe acute respiratory syndrome coronavirus 2 (SARS-CoV-2) began in the city of Wuhan in China which spread to over 30 countries of the world [1]. In the last 25 years, notable highly infectious respiratory viruses with pandemic potentials has emerged and remerged. Notable of which is the influenza virus that issued a global alert in 1918, 1957, 1968, 2003, and 2019 causing severe acute respiratory diseases [2]. The Novel SARS-CoV-2 outbreak resulted in globally, as of 2:33 pm CEST, 17 May 2020, there have been 4,525,497 confirmed cases of COVID-19, including 307,395 deaths, reported to WHO with substantial economic impact. Since then several other viral respiratory pathogens have emerged including Middle East respiratory syndrome coronavirus (MERS-CoV), adenovirus-14, and virulent strains of influenza viruses. Soon after the discovery of SARS, new coronaviruses NL63 and HKU1 were identified [2, 3]. The emergence of 35 different respiratory viruses underscores the epidemic potential and overall threat to global health security. Severity caused by Novel SARS-CoV-2 has been recognized as a global public health security threat [3]. Many African countries are not prepared for the Novel SARS-CoV-2 outbreak due to poor and weak healthcare system, poor surveillance and response system, as well as inadequate and overstretched health facilities and services established higher risk of Novel SARS-CoV-2 importation from Europe to Africa than china importation, comparing rapid spread of the virus in selected sub-Sahara countries than in European countries. Some African countries have developed capacity to respond to the outbreak as at 11 May, 2020, a total of 13,814 confirmed cases and 747 deaths from Novel SARS-CoV-2 have been documented in Africa [2].

Genome sequence associated with Middle East respiratory syndrome (MERS) and human severe acute respiratory syndrome (SARS) has been systematically analyzed to be linked to beta bat coronavirus [4], WHO officially named the virus "SARS-CoV-2" although its origin is still been investigated which suggests human to human transmission could be through wild animals been sold illegally at a wholesale seafood market in Wuhan [5]. In this review, we summarized the latest research progress on the transmission mode dynamics and viral shedding in provide direction for Isolation protocol. The transmissibility of SARS-CoV-2 is represented by the reproduction number (R0) which is the average number of new infections generated by an infectious person in a totally naïve population [6].

2. SARS CoV-2 viral genome and key virulence factors

A 29.9 kb weight of genome structure which are key virulence factors where profile from SARS CoV-2 patients in Wuhan market in China [7]. While SARS-CoV and MERS-CoV have positive-sense RNA genomes of 27.9 and 30.1 kb, respectively [8]. MERS-CoV and SARS-CoV 2 are made up of 27.9 and 30.1 kb RNA genomes positive-sense with 6–11 variable opening frame (ORFs) [9]. The first ORF (ORF1a/b) location translate pp1a and pp1ab polyprotein which is made up of two-third RNA viral genome encoded in the 16 non-structural proteins (NSP) while other encasement are of structural and accessory protein ORFs. Other essential viral structural proteins include; nucleocapsid (N) protein, matrix (M) protein, small envelope (E) protein and spike (S) glycoprotein [10], It was established by Wu et al. that several accessory proteins interfere with innate immune response of the host [7].

3. Epidemiology—origin, Reservoirs and transmission dynamics of SARS-CoV-2

The SARS-CoV-2 is positive-sense RNA virus from the family of β-coronavirus with a non-segmented envelope belonging to the subfamily of Orthocoronavirinae and sarbecovirus subgenus [4]. α-/β-/γ-/δ-CoV. α- and β-CoV are the four genera of coronavirus that infect mammals. Birds are infected by γ- and δ-CoV genera. Six genera have been identified to cause mild respiratory tract infection in humans, they include; HCoV-OC43, β-CoVs HCoV-HKU1, HCoV-NL63 and HCoV-229E while fatal respiratory tract infection in human is caused by SARS-CoV, β-CoVs and MERS-CoV. There is similarity in homology genome sequence between SARS-CoV-2 and bat CoV RaTG13 with 96.2% identity. Evolutionary analysis suggest SARS CoV-2 is transmitted to humans from bat as intermediate host with special viral tropism to angiotensin-converting enzyme 2 (ACE2) receptors. On December 12, 2019, an epidemic of unknown origin broke out in Wuhan province of China causing acute respiratory tract infection in human population. Source of infection was traced to seafood market. Studies suggest Bat might be the reservoir host of SARS-CoV-2 [6, 11]. Phylogenetic analysis of protein sequence alignment reveals intermediate host such as turtles and pangolin as sources of human to human transmission and also implicated in nosocomial transmission seen within health care workers as revealed on 14 February 2020 by National Health Commission of China [12].

4. Detection of SARS-CoV-2 in different types of clinical Specimens

Reveals SARS-CoV-2 was detected in 205 patients at multiple sites with lower respiratory tract samples, importantly the RNA virus has been detected in feces which imply SARS-CoV-2 may be transmitted by the fecal route. A small percentage of blood samples had positive PCR test results, suggesting that infection sometimes may be systemic. Transmission of the virus by respiratory and extra respiratory routes may help explain the rapid spread of disease. In addition, testing of specimens from multiple sites may improve the sensitivity and reduce false-negative test results. Two smaller studies reported the presence of SARS-CoV-2 in anal or oral swabs and blood from 16 patients in Hubei Province, 3 and viral load in throat swabs and sputum from17 confirmed cases.

Retrospectively identified a convenience sample of patients admitted to Beijing Ditan Hospital, Capital Medical University, with a diagnosis of COVID-19 and paired RT-qPCR testing of pharyngeal swabs with either sputum or feces samples. A diagnosis of COVID-19 required at least 2 RT-qPCR–positive pharyngeal swabs, and patients underwent treatments as well as initial and follow-up testing of pharyngeal, sputum, or fecal samples at the discretion of treating clinicians. Hospital discharge required meeting four criteria: afebrile for more than 3 days, resolution of respiratory symptoms, substantial improvement of chest computed tomographic findings, and two consecutive negative RTqPCR tests for SARS-CoV-2 in respiratory samples obtained at least 24 hours apart [13]. We report the findings of patients with at least one initial or follow-up RT-qPCR positive sputum or fecal sample obtained within 24 hours of a follow-up negative. RT-qPCR pharyngeal sample. The RT-qPCR assay targeted the open reading frame 1ab (ORF1ab) region and nucleoprotein (N) gene with a negative control. A cycle threshold value of 37 or less was interpreted as positive for SARS-CoV-2, according to Chinese national guidelines. Among 133 patients admitted with COVID-19 from 20 January to 27 February 2020, we identified 22 with an initial or follow-up positive sputum or fecal samples paired with a follow-up negative pharyngeal sample. Of these patients, 18 were aged 15–65 years, and 4 were children; 14 were male; and 11 had a history of either travel to or exposure to an individual returning from Hubei Province in the past month. Fever was the most common initial onset symptom.

5. Clinical characteristics, complications and clinical outcomes

Direct contact, respiratory secretions and droplets from respiratory tract are emerging rout of SARS-CoV-2 spread [10]; SARS-CoV-2 was isolated from fecal samples of severe pneumonia patients at Sun Yat-Sen University, Guangdong, China on February 2020, Zhang et al. [14]. ACE2 protein abundance on lung alveolar epithelial cells and enterocytes of small intestine has been discovered [15], which may reveal broad understanding of the routes of infection and disease. Epidemiological investigation reveals signs and symptoms to SARS-CoV-2 becomes manifest between 1 and 14 days, mostly 3–7 days suggesting SARS-CoV-2 can be contagious during a latency period. Elderly and individuals with underlying diseases are at risk of acquiring SARS-CoV-2. A median age of 47–59 years and 41.9–45.7% of patients were females [10, 12, 16]. Comorbidities associated with SARS-CoV-2 in adult might lead to flu like symptoms, malaise, cough which might lead to respiratory failure, distress syndrome and even dead. SAR-CoV-2 patients had good clinical outcome except for few that have associated comorbidities. As at March 1st 2020, there are 79,968 confirmed cases with severe cases totaling 14,475 (18.1%) and 2873 deaths (3.5%) from the China mainland as reported by the WHO [2]. liver dysfunction, acute cardiac injury, Arrhythmia, acute respiratory distress syndrome (ARDS), acute kidney injury are among associated complication [16]. The severity of the disease is associated with poor clinical outcome mostly seen among the elderly which progress faster with dead mostly seen among people aged 65 years [16, 17].

6. Conclusion

The global outbreak of SARS-CoV-2 is across 85 countries. Our study revealed that person to person transmission within family cluster or Nosocomial infection is possible in setting where precautions such as personal hygiene, social distancing and the use of personal protective equipment are not adhered to. Clinicians should be aware of clinical history of contact patients to enable them promptly identify in order to curb further spreading in hospital and family cluster.

Our recommendation will be for adoption of National Guideline that will reveal epidemiological exposure history as an important reference point for identifying the source of infection and strengthened protection, and isolation measures. Close contacts to confirm Cases should be included highly Suspected Cases during Incubation period of Confirmed Cases. Availability of high sensitive rapid diagnostic reagents for Novel SARS-CoV-2 should be accelerated in order to facilitate community testing.

Author details

Adamu Ishaku Akyala
Department of Microbiology, Faculty of Natural and Applied Sciences,
Nasarawa State University, Keffi, Nasarawa State, Nigeria

*Address all correspondence to: i.adamu@erasmusmc.nl

References

[1] Lu R, Zhao X, Li J, Niu P, Yang B, Wu H, et al. Genomic characterisation and epidemiology of 2019 novel coronavirus: Implications for virus origins and receptor binding. Lancet. 2020;**395**(10224):565-574

[2] WHO. Coronavirus Disease (COVID-2019) Situation Reports. 2020. Available from: https:// www.who.int/emergencies/diseases/novel-coronavirus-2019/situationreports [Accessed: 05 March 2020]

[3] Riou J, Althaus CL. Pattern of early human-to-human transmission of Wuhan 2019 novel coronavirus (2019-nCoV), December 2019 to January 2020. Euro Surveillance. 2020;**25**(4):2000058. DOI: 10.2807/1560-7917.ES.2020.25.4.2000058

[4] Zhu N, Zhang D, Wang W, Li X, Yang B, Song J, et al. A novel coronavirus from patients with pneumonia in China, 2019. The New England Journal of Medicine. 2020;**382**(8):727-733

[5] Zhou P, Yang XL, Wang XG, Hu B, Zhang L, Zhang W, et al. A pneumonia outbreak associated with a new coronavirus of probable bat origin. Nature. 2020;**579**(7798):270-273. DOI: 10.1038/s41586-020-2012-7

[6] Giovanetti M, Benvenuto D, Angeletti S, Ciccozzi M. The first two cases of 2019-nCoV in Italy: Where they come from? Journal of Medical Virology. **2020**:1-4. DOI: 10.1002/jmv.25699 [Epub ahead of print]

[7] Wu F, Zhao S, Yu B, Chen YM, Wang W, Song ZG, et al. A new coronavirus associated with human respiratory disease in China. Nature. 2020. DOI: 10.1038/s41586-020-2008-3 [Epub ahead of print]

[8] de Wit E, van Doremalen N, Falzarano D, Munster VJ. SARS and MERS: Recent insights into emerging coronaviruses. Nature Reviews. Microbiology. 2016;**14**(8):523-534

[9] Song Z, Xu Y, Bao L, Zhang L, Yu P, Qu Y, et al. From SARS to MERS, thrusting coronaviruses into the spotlight. Viruses. 2019;**11**(1):E59. DOI: 10.3390/v11010059

[10] Cui J, Li F, Shi ZL. Origin and evolution of pathogenic coronaviruses. Nature Reviews. Microbiology. 2019;**17**(3):181-192

[11] Paraskevis D, Kostaki EG, Magiorkinis G, Panayiotakopoulos G, Sourvinos G, Tsiodras S. Full-genome evolutionary analysis of the novel corona virus (2019-nCoV) rejects the hypothesis of emergence as a result of a recent recombination event. Infection, Genetics and Evolution. 2020;**79**:104212

[12] Guan WJ, Ni ZY, Hu Y, Liang WH, Ou CQ, He JX, et al. Clinical characteristics of coronavirus disease 2019 in China. The New England Journal of Medicine. 2020;**382**(18):1708-1720. DOI: 10.1056/NEJMoa2002032

[13] Li W, Shi Z, Yu M, Ren W, Smith C, Epstein JH, et al. Bats are natural reservoirs of SARS-like coronaviruses. Science. 2005;**310**(5748):676-679

[14] Angeletti S, Benvenuto D, Bianchi M, Giovanetti M, Pascarella S, Ciccozzi M. COVID-2019: The role of the nsp2 and nsp3 in its pathogenesis. Journal of Medical Virology. 2020;**92**(6):584-588. https://doi.org/10.1002/jmv.25719

[15] Zhang L, Shen FM, Chen F, Lin Z. Origin and evolution of the 2019 novel coronavirus. Clinical Infectious Diseases. 2020. DOI: 10.1093/cid/ciaa112 [Epub ahead of print]

[16] Tang X, Wu C, Li X, Song Y, Yao X, Wu X, et al. On the origin and continuing evolution of SARS-CoV-2. National Science Review. 2020;**7**(6):1012-1023. DOI: 10.1093/nsr/nwaa036

[17] Jia HP, Look DC, Shi L, Hickey M, Pewe L, Netland J, et al. ACE2 receptor expression and severe acute respiratory syndrome coronavirus infection depend on differentiation of human airway epithelia. Journal of Virology. 2005;**79**(23):14614-14621

Chapter 2

Assembling an Anti-COVID-19 Artillery in the Battle against the New Coronavirus

Chanda Siddoo-Atwal

Abstract

The panic and confusion surrounding the pandemic caused by the novel coronavirus requires a systematic study of the disease (COVID-19) and the arsenal of weapons available to the biochemist in the fight against infection. When developing a particularly bad flu in January 2020 while in India after the visit of a friend, who had just travelled back from Wuhan (China), it gave me an early opportunity to study the tricky diagnosis of this dreaded disease first-hand. The somewhat unusual symptoms and a lingering weakness and malaise for months suggested that it was no ordinary influenza virus. Since that time, a baffling number of disparate symptoms have been ascribed to COVID-19 infection including respiratory, gastrointestinal, circulatory, urinary tract and nerve dysfunction that have even resulted in multi-organ failure in some cases. Naturally, an array of risk factors have also been identified ranging from age, sex, obesity, diabetes, and hypertension to cigarette smoking that can increase mortality rate dramatically. In the intervening period, much research has appeared on biochemical compounds that may help to prevent this infection and, possibly, aid in patient recovery. Among these bioactive molecules are certain anti-inflammatory substances such as vitamin D, zinc, chloroquine, soy isoflavones like genistein, and glycyrrhizic acid, some of which may be successful in attacking different biochemical processes of the new coronavirus and disarming its deadly artillery against the human host. In a few instances, the viral processes that are inhibited by these chemicals are essential for the replication and reproduction of this RNA virus thereby striking a lethal blow to its machinery. Thus, taken together, these compounds may form a worthy arsenal against a formidable foe in the absence of an effective vaccine, and, especially, if relapse or re-infection proves to be a common occurrence in recovered COVID-19 patients.

Keywords: novel coronavirus, COVID-19, influenza, obesity, diabetes, hypertension, cigarette smoke, vitamin D, zinc, chloroquine/quinine, soy isoflavones, genistein, glycyrrhizic acid, RNA virus replication

1. Life cycle of the novel coronavirus

Coronaviruses are large, enveloped, single-stranded, positive-sense RNA viruses with a genome of approximately 30 kilobases in length. The genus *Coronavirus* belongs to the family *Coronaviridae* in the order *Nidovirales*. They are classified into three groups. Group 1 contains various mammalian viruses including porcine epidemic diarrhea virus, porcine transmissible gastroenteritis virus, and human coronaviruses 229E and NL63. Group 2 includes canine respiratory coronavirus among other mammalian viruses and human coronavirus OC43. Human severe acute respiratory syndrome coronavirus (SARS-CoV-1) is considered a distant relative of this group. Group 3 contains solely avian coronaviruses. Human coronaviruses (HCoVs) cause respiratory infections, mainly, but gastroenteritis and neurological disorders may also occur. So far, at least seven human coronaviruses have been described including SARS-CoV-2, which was just sequenced in 2020, and two of these coronaviruses (OC43 and 229E) are responsible for 10–30% of all common colds. HCoV-HKU1 is mostly associated with bronchiolitis and pneumonia [1–3].

The gross life cycle of the SARS-CoV-1 has been observed in Vero E6 cells (African green monkey kidney cells) following inoculation with the virus under an electron microscope. The SARS-CoV-1 enters the cells through membrane fusion. Then, the nucleocapsids are assembled in the rough endoplasmic reticulum (RER) and mature by budding into the smoothe vesicles derived from the Golgi apparatus. Finally, the smoothe vesicles fuse with the cell membrane and the mature virus particles are released [4]. SARS-CoV-2 displays a similar life cycle.

Recent molecular studies have revealed that in order to facilitate entry of the virus into a human cell, the "S" spike surface glycoprotein of SARS-CoV-2 binds to the angiotensin-converting enzyme 2 (ACE-2) cellular receptor. Binding of the virus occurs via the S1 subunit of the S protein to a receptor and entry requires S protein priming by the cellular serine protease in order to allow fusing together of viral and cell membranes, a process which is initiated by the S2 subunit [5]. Following the fusion of viral and plasma membranes, the virus RNA undergoes transcription and replication inside the cell cytoplasm. Viral proteins are synthesized and the new RNA genomes are assembled and packaged in the endoplasmic reticulum, in the Golgi apparatus, and in the endoplasmic reticulum-Golgi intermediate compartment prior to virion release in vesicles. In fact, the S protein of SARS-CoV-2 binds to ACE-2 receptors with an approximately 10–20 fold higher affinity than that of SARS-CoV-1 and this added feature may aid in the efficient spread of SARS-CoV-2 among human populations. However, SARS-CoV-2 does not employ the other usual CoV receptors such as aminopeptidase N and dipeptidyl peptidase 4 to enter human cells [6].

ACE-2 is a membrane-associated aminopeptidase that converts angiotensin II to angiotensin 1–7 and plays a role in the cleavage of peptides [3]. Expression of ACE-2 in human tissues correlates with known sites of SARS-CoV-1 infection including lungs (particularly airway epithelia), heart, kidneys, small intestine, testes, and vascular endothelia [7]. These same tissues also overlap with the sites of SARS-CoV-2 infection in humans due to ACE-2 receptor availability.

2. A personal experience

On a personal note, as a biochemist, I have been following every bit of new research on any chemical compound that might successfully combat the virus. Around January 6th, 2020, I developed a very bad flu while in India after meeting with a friend who had just travelled to Wuhan in China. Overnight, I got a sore-throat that lasted a few days followed by a severe head cold with sinus congestion and mucous and, finally, it developed into a dry cough. During this debilitating flu, I also had some loose bowel movements with mucous. In the aftermath of the flu that lasted around 14 days, I was plagued with dizziness and weakness for two more weeks.

Although we had heard of the novel coronavirus in China, there was no reason to believe that was what I had just experienced since there had been no unusual respiratory distress. So, it did not seem to overlap with the pneumonia-like symptoms of the new coronavirus from China. Moreover, the friend had returned at the beginning of January and, as far as we knew at that time, the virus had only appeared in

December. Therefore, it seemed unlikely that the friend had been exposed to any infected individuals while in China. Furthermore, the traveller from China never became sick (although one other person who attended the same meeting as myself developed a very bad flu within two weeks of coming in contact with this person). At the same time, there are also many seasonal flus like swine flu (H1N1) that are endemic in India, so there was no reason to consider that it was a coronavirus infection. Finally, the only medicines I took initially were some herbal Ayurvedic cold remedies mainly with a licorice-root base (a potent anti-inflammatory), aspirin at night, and an electrolyte solution to prevent dehydration from diarrhea. When I had a relapse of the gastrointestinal symptoms in March including stomach pain after I returned to Canada, a course of azithromycin helped to resolve the symptoms.

However, it was only when the weakness and malaise persisted for 3–4 months after the initial illness and new data started to emerge about the differing patterns of COVID-19 infection, that I started to consider another possible cause. Firstly, all my symptoms were consistent with the disparate effects of the novel coronavirus including the lingering apathy. Secondly, it became apparent that the new coronavirus had appeared in Wuhan some time before December. Thirdly, unlike other flu viruses, the phenomenon of asymptomatic spreaders became widely known. So, now, even though I had not been tested for the new virus or COVID-19 antibodies, I started to suspect that I could have experienced a form of coronavirus infection.

Finally, I had my COVID-19 test in August 2020 and, although it was negative, it did not preclude the possibility that I had the disease in January 2020 and that my body had formed and shed antibodies to the novel coronavirus (antibody testing was also negative). Since it is not known exactly how long antibodies persist following infection, even these may not be detected after a certain recovery period (there are recent reports antibodies decline after three months). Studies in rhesus monkeys show that re-infection does not occur in the recovered macaques up to 28 days after initial infection [8]. Nevertheless, prolonged inflammation and reports of re-infection in recovered humans are a surprising aspect of this virus. In my case, one additional negative C-reactive protein (inflammatory marker) test decisively clinched the matter.

3. Top COVID-19 risk factors

3.1 Internal risk factors

Some scientists have opined that COVID-19 is highly contagious and highly lethal to a small subset of the population, while it produces milder symptoms in most people. Although the SARS-CoV-2 virus infects people of all ages, the World Health Organization (WHO) has determined that the evidence to date suggests that older adults and adults with underlying medical conditions are at a higher risk of developing severe COVID-19 disease [9].

One large study out of New York State seems to indicate that obesity, high blood pressure, and diabetes are strong risk factors for COVID-19 [10]. It has also been observed that cardiovascular disease and respiratory diseases could greatly affect the prognosis [11]. In fact, in an interesting German study involving autopsies on 12 COVID-19 patients the results revealed that coronary heart disease and asthma were common comorbid conditions in 50% of the deceased [12].

Other research suggests that cancer patients are more vulnerable to COVID-19 infection. A multicenter study showed that patients with cancer had higher risks in all severe outcomes of the disease tested. Hematologic cancer, lung cancer, or metastatic cancer (stage IV) cases experienced the highest frequency of severe events, while nonmetastatic cancer cases experienced similar frequencies to patients without cancer. Moreover, cancer patients who received surgery had higher risks of severe events than patients without cancer or those who underwent radiotherapy [13].

In addition, a surprising gender disparity appears to be present in relation to SARS-CoV-2 infection. Statistics from Australia, Belgium, Germany, Italy, the Netherlands, South Korea, Spain, the U.K and the US reveal that mortality rates from the virus are significantly higher in infected males than in infected females. In New York, approximately 60% of COVID-19-related deaths occurred in men. This may partly reflect biological characteristics since women produce stronger immune responses than men and are physically better at warding off viral and other types of infections. Nevertheless, biochemical differences in sex hormones are also likely to play a role in determining this dichotomy [14] and certain researchers have suggested it may be due to the presence of ACE-2 receptors in the testicles [15].

In the largest Chinese study to date assessing severity of coronavirus infection in smokers, it was found that higher percentages of current and former smokers needed ICU support or mechanical ventilation. Higher percentages of smokers among the severe cases also died [16]. Therefore, ultimately, the risk of any one individual is determined by the number of risk factors they display. For example, a ninety year old male smoker with diabetes and hypertension displaying five risk factors (age, gender, smoke inhalation, high blood pressure, and diabetes) would have an extremely high risk of contracting a terminal case of COVID-19.

However, genetic risk factors as a result of ethnic origin can only be considered once all these other significant risk factors have been taken into consideration. So far, despite attempts by various institutions to prove an ethnic link to COVID-19 infection, there is no compelling evidence to suggest that any one human group is genetically more susceptible to the novel coronavirus than any other beyond mitigating factors such as socioeconomic status or environmental conditions [17]. In order to establish a true genetic component, rigorous genetic testing must be undertaken to identify predisposing genes in susceptible ethnic groups. Prior to gene isolation and identification of a specific genetic polymorphism, a biochemical reaction resulting in a higher percentage of the disease is often demonstrated in a particular human population. As an example, the human sunburn cycle in response to UVA/B radiation only occurs in a minority of people with fair skin; however, most people simply tan when they are exposed to sunlight. In fact, these represent two separate physiological processes (burning and tanning). The former condition, scientific sunburn as a result of the human sunburn cycle, is mostly due to a genetic polymorphism involving the expression of very low levels of melanin in human skin since it can be corrected by wearing a sunscreen containing black sesame melanin [50 mg/ml] in a zinc oxide cream base [7.5%] [18–20]. It is also correlated with a high risk for skin cancer. Nonetheless, there may be other genetic factors like differences in DNA repair enzyme activity which can contribute to this unusual trait in certain individuals, as well [21].

Simultaneously, a surprising recent genetic association study has revealed that a major genetic risk factor for severe COVID-19 in humans may actually be inherited from Neanderthals. Outside the continent of Africa (0.3%), modern humans have inherited significantly more genetic material from other hominid species including Neanderthals (approximately 2%) and Denisovans [22]. Europeans and South Asians appear to have the greatest complement of Vindija Neanderthal genes from Croatia and a gene cluster on chromosome 3 inherited from this species has been identified as a risk locus for respiratory failure after infection with SARS-CoV-2. Among certain South Asian populations, up to 50% can carry at least one copy of this risk haplotype and the highest carrier frequency occurs in Bangladesh where 63% of the population carries it. In the UK it has been reported that individuals of Bangladeshi origin have roughly a two times higher risk of dying from COVID-19 than people of other nationalities [23].

3.2 External risk factors

Interestingly, there are high levels of air pollution in the two regions of China and Northern Italy that were hardest hit by the virus suggesting that environmental conditions can have an impact on the infectiousness of the disease [24]. Italian researchers have recently proposed an association between higher mortality rates in Northern Italy and peaks of particulate matter concentrations in this region. The most polluted northern provinces of Italy were found to have more infection cases than the less polluted southern provinces and this correlated well with ambient particulate matter concentrations that often exceeded the legal limit in these areas. All data for this study was collected prior to the lockdown [25]. Surprisingly, further research by the same group demonstrated that SARS-CoV-2 RNA was present on outdoor airborne particulate matter that was collected from an industrial site in Bergamo, Italy. This evidence suggests that, under the right atmospheric conditions, SARS-CoV-2 could create clusters with particulate matter and enhance persistence of the virus in the atmosphere by facilitating its capacity for diffusion. However, the vitality and virulence of the coronavirus diffused via this method remain to be confirmed [26].

This could have been a significant factor in the spread of the coronavirus in highly polluted and populated cities like Mumbai, India. Social conditions such as crowding in slums have also been considered contributory to dispersal of the virus in developing countries like Brazil and India. Proximity to infected individuals increases the risk of person-to-person transmission since the SARS-CoV-2 virus is spread mainly by respiratory droplets, but can be aerosolized, too [3].

No matter how healthy an individual may be, the more exposure they have to a particular virus, the greater risk they have of contracting the disease. The greater the number of particles of the virus one is exposed to, the greater the chance that they will overwhelm the body and immune responses. This is the reason that young doctors and other frontline healthcare workers are getting serious cases of COVID-19 and dying at a higher frequency than the general population.

View of Downtown Mumbai – December 2019

View of Mumbai Harbour – December 2019

4. An array of symptoms and complications

In general, COVID-19 infection is associated with the increased production of pro-inflammatory cytokines, C-reactive protein, increased risk of pneumonia, sepsis, acute respiratory distress syndrome, and heart failure [24]. In fact, a cluster of unexplained pneumonia cases were first reported in Wuhan, China in late December 2019. A few days later, the cause of this pneumonia was identified as a new member of the coronavirus family. Since then, the virus has spread throughout China and precipitated a global pandemic [6].

Early reports from China suggested the most common symptoms of COVID-19 infection were fever (88%) and dry cough (67.7%). Rhinorrhea (4.9%) and gastrointestinal symptoms (diarrhea 4–14%) were less common. At the same time, a majority of patients (81%) had only mild symptoms (no pneumonia or mild pneumonia). Among patients with more pronounced symptoms, 14% experienced severe symptoms while 5% were critically ill with respiratory failure, septic shock, or multiorgan dysfunction or failure [3].

Although the novel coronavirus preferentially infects cells in the respiratory tract, autopsy results from Germany showed that it can be detected in multiple organs. The highest levels of the virus were detected in the lungs and the respiratory tract, while lower levels were usually present in the heart, liver, brain, kidneys, and spleen. This data suggests that SARS-CoV-2 may spread via the bloodstream and infect other organs. It also appears that COVID-19 may predispose patients to venous thromboembolism in several different ways including via endothelial dysfunction and promotion of a procoagulatory state by tissue factor pathway activation. High plasma levels of proinflammatory cytokines were observed in a small subset of patients with severe COVID-19 and, therefore, direct activation of the coagulation cascade by a cytokine storm is also plausible [12].

In one study, it was found that 22% of critically ill patients experienced myocardial injury from the infection [3]. In another study, the incidence of thrombotic complications in ICU patients with COVID-19 infections was reported to be 31%. It was concluded that COVID-19 may predispose to both venous and arterial thromboembolism due to excessive inflammation, hypoxia, immobilization, and diffuse intravascular coagulation [27].

In addition, the COVID-19 pandemic is surprisingly associated with neurological symptoms and complications including anosmia, hypogeusia, seizures, and stroke. Although statistics are not widely available at this point, the clinical course of COVID-19 is most severe in elderly male patients with comorbidities such as hypertension, diabetes, heart disease, and obesity which are all risk factors for stroke. There appears to be hypercoagulability associated with COVID-19 as a result of a "sepsis-induced coagulopathy" that may be a predisposing factor due to the formation of blood clots in the body [28]. COVID-19 complications in the brain can include delirium, inflammation, and encephalitis. A new study from UCL suggests that serious problems can occur even in individuals with mild cases of the virus [29].

A temporary loss of smell (anosmia) can be a consistent indicator of COVID-19 infection. An interesting finding is that the virus seems to change the sense of smell in patients by infecting and affecting the function of non-neural cells that support olfactory neurons. However, the neurons themselves do not appear to be infected as they do not express ACE-2 receptors [30].

Diabetes is already known to be a risk factor for COVID-19 and diabetics are more likely to die from the disease. Now, mounting evidence suggests that not only does diabetes make patients more vulnerable to the novel coronavirus, but the virus may actually trigger diabetes in some. Preliminary tissue studies indicate that the virus may act by damaging insulin-producing cells in the pancreas of affected individuals [31].

Even though, initially, children were thought to be unaffected by the novel coronavirus, a cluster of children with hyperinflammatory shock and features similar to Kawasaki disease and toxic shock syndrome was first reported in England. This hyperinflammatory condition could lead to severe illness, multiorgan failure, and even death in extreme cases. New reports out of the UK and US suggest that symptoms in young children (mainly toddler to elementary school age) can include inflammation of the blood vessels and coronary arteries. Almost all these pediatric cases had positive SARS-CoV-2 test results. As a result, this illness has been termed COVID-19-associated multisystem inflammatory syndrome [32].

5. The scope of coronavirus vaccines

Historically, there is no doubt that vaccines have provided a tremendous tool against infection for a variety of microbes including those causing small pox, tetanus, typhoid, cholera, and polio. Vaccines are an effective way for a population to achieve herd immunity (the concept that a pandemic will end once 60–70% of people become immune to any particular virus or microorganism). However, more recently, there are instances in which the production of viral vaccines has not been so successful as in the case of human immunodeficiency virus (HIV) and human coronaviruses (HCoVs) possibly due to their complex genomes. Virologists and immunologists maintain that it takes up to ten years to prepare a really good vaccine that has been properly tested. In fact, some of these specialists are skeptical about the race to find the first vaccine for the novel coronavirus within one year and often strike a cautionary note.

There are a number of things to consider in connection with a SARS-CoV-2 vaccine. Firstly, even if a safe and effective vaccine is made against the novel coronavirus, it may not be widely available in time to make a significant difference to the pandemic. Secondly, no successful vaccine against *any* coronavirus has been produced so far despite seventeen years of research. Moreover in March, the British Society for Immunology published an open letter stating that it is unknown whether this virus will induce long-term immunity in affected individuals as other related viruses do not [33]. Thirdly, certain vaccines can protect against a disease, but not against infection, so vaccinated individuals could potentially become asymptomatic carriers of SARS-CoV-2. Fourthly, some vaccines developed against SARS-CoV-1 (a close viral relative of SARS-CoV-2) actually exacerbated the disease in mice. Fifthly, although the easiest way to make a vaccine is to inactivate the pathogen, there are new vaccines in current trials based on RNA from coronaviruses or other RNA viruses that have never before been approved or tested in humans. Therefore, there could conceivably be unintended or irreversible consequences. Finally, at least one of the novel coronavirus vaccines approved for clinical trials so far has caused severe adverse events in three of eight healthy, young individuals that were tested [34] and other trials have been suspended. Unfortunately, the contamination of vaccines which are mass produced for a burgeoning human population also seems to be a potential problem and an ideal tool for rival countries to conduct biological warfare upon each other. Oral or nasal vaccines may be safer in this respect [35]. In addition, there is a physical limit to the number of vaccines a person can safely receive as new and deadlier viruses arise in the environment.

6. Biochemical effects of special supplements

6.1 Vitamin D

The action of UVB radiation striking and reacting thermally with 7-dehydrocholesterol in human skin results in the production of Vitamin D3 in the human body. This form of Vitamin D is converted to the hormonal metabolite, calcitriol, in a set of biochemical reactions in the liver, kidneys, and other organs as required. Then, calcitriol binds with the nuclear vitamin D receptor, which is a DNA binding protein, that interacts directly with regulatory sequences near target genes and affects their transcriptional output.

Vitamin D also enhances cellular innate immunity partly through the induction of antimicrobial peptides, including human cathelicidin, and, defensins. Cathelicidins exhibit direct antimicrobial activities against a spectrum of microbes including many types of bacteria, enveloped and nonenveloped viruses, and fungi. The main action of these host-derived peptides is to kill the invading pathogens by perturbing their cell membranes. Moreover, vitamin D is effective in reducing concentrations of pro-inflammatory cytokines that produce the inflammation that injures the lining of the lungs leading to pneumonia during viral infections like COVID-19 and increasing concentrations of anti-inflammatory cytokines [24].

According to a recent clinical study with a large sample size taken from different countries around the world, vitamin D supplements were found to protect against respiratory tract infections including colds and influenza. The most benefit was observed in patients who were very vitamin D deficient. This protective effect is likely provided by the capacity of vitamin D to boost levels of antimicrobial peptides in the lungs [36].

Vitamin D deficiency is a world-wide problem, but is particularly pronounced in the elderly, who are at greatest risk of contracting severe COVID-19 infection. The release of pro-inflammatory cytokines is one of the major causative factors in serious COVID-19 infections. However, vitamin D modulates their presence in the body by preventing macrophages from releasing too many inflammatory cytokines and chemokines. Calcitriol has also been found to exert an influence on ACE-2 receptors. Thus, it is not surprising that vitamin D deficiency has been correlated with COVID-19 cases and an increased risk of mortality in a European study [37].

6.2 Zinc

RNA synthesis occurs in the life cycle of the SARS-CoV-1 virus in order to reproduce its genetic material and is catalyzed by an RNA-dependent RNA polymerase, which is the core enzyme of a multiprotein replication/transcription complex. In the case of SARS-CoV-1, an excess of intracellular zinc ions has been found to efficiently inhibit the RNA-synthesizing activity of this replication and transcription multiprotein. Enzymatic studies *in vitro* have revealed that zinc directly blocks the activity of the RNA polymerase by inhibiting elongation and reducing template binding. This RNA polymerase core, which is a central component of the coronaviral replication/transcription machinery, is well conserved among the members of the coronavirus family including SARS-CoV-2 [38, 39]. Therefore, it is quite possible that zinc treatment would have a similar biochemical effect on SARS-CoV-2 and interfere with its ability to replicate.

Since current research indicates that the mineral, zinc, can inhibit the replication of coronavirus and a variety of other RNA viruses in cell culture, it has become a potentially important and interesting supplement to study at this time. In the human body, zinc performs a variety of vital antioxidant functions and is required for maintaining good health. Inside the cell, the harmful effects of free radicals are balanced by the action of antioxidant enzymes (such as copper-zinc superoxide dismutase) and non-enzymatic antioxidants (such as metallothioneins). As zinc cannot pass easily through membranes, zinc-transporting proteins, ZIPs (Zrt-Irt-like protein or Zinc Iron permease) and ZnTs (Zinc transporters) help to facilitate this process. Metallothionein also aids in the regulation of zinc levels and the distribution of this metal in the extracellular space. The presence of zinc within the cell causes an increase in metallothionein, which is the major zinc-binding protein, and together they form a thermodynamically stable complex [40, 41]. Thus, low risk ways of increasing zinc bioavailability in the body can be safely considered.

In rats, rice fortified with zinc oxide or zinc carbonate is a feasible vehicle for zinc absorption, although zinc oxide displays lower bioavailability than zinc carbonate [42]. In young adults, zinc absorption from supplemental zinc citrate is comparable with that from zinc gluconate, but higher than from zinc oxide [43]. It is already known that zinc can be absorbed from topical (non-nano) zinc oxide by human skin in small quantities (nano forms of zinc oxide are not associated with significant zinc absorption) [44]. One of our recent studies suggests that zinc is absorbed by the human body from our suncare products (all with the same basic formula containing a medicinal form of zinc oxide) in sufficiently large quantities with regular use [45].

So, recently, when our company received an inquiry from Health Canada regarding any innovations that may benefit Canadian health workers at this critical time during the novel coronavirus pandemic, the answer was that we do have a product that may be useful to medical professionals and health workers in the field. It is a natural, award-winning suncare product specially formulated to block apoptotic sunburn (Skin Protector Plus). Its active ingredient is a non-nano, medicinal form of zinc oxide. The novel thing about this product is that it appears to be an efficient delivery system for boosting zinc levels in the whole body in a relatively short period of time. There is no toxicity associated with this product due to the use of high grade zinc oxide and natural ingredients. Since it is so safe and contains no harsh chemicals (already tested on human volunteers), no pre-clinical trials would be required to test its efficacy in protecting subjects from COVID-19 in a clinical study. The objective of such a comprehensive study would be to test and confirm the hypothesis outlined above, *in vivo*; namely, if maximum zinc levels are maintained in the human body via percutaneous zinc absorption from a topically applied zinc oxide cream, then it may provide one suitable defense against SARS-CoV-2 infection. Although oral supplementation is also an option, this type of topical application on the surface of the skin may be a faster method of ensuring even zinc distribution throughout the body and delivery to the various potential points of viral entry. Moreover, it may actually provide a physical barrier or blockade against entrance of the virus into the body by allowing suffusion and accumulation of zinc pools directly beneath the skin.

6.3 Chloroquine/quinine

Quinine, an alkaloid derived from the bark of the cinchona tree, is most commonly found in South America, Central America, the islands of the Caribbean, and parts of the western coast of Africa. It is an important antimalarial drug and a synthetic form with a similar mode of action is known as chloroquine [46]. Chloroquine has been reported to inhibit the SARS-CoV-1 virus in infected cell cultures *in vitro* at doses equivalent to those used in the treatment of acute malaria in humans. Its antiviral effect appears to depend on the fact that chloroquine is a weak base that increases the pH of acidic vesicles when added extracellularly. The nonprotonated portion of chloroquine enters the cell where it becomes protonated and concentrated in acidic, low-pH organelles such as endosomes, Golgi vesicles, and lysosomes. The subsequent antiviral activity of the chloroquine depends partly on the extent to which a particular virus utilizes endosomes for entry into the cell [47]. In addition, this drug appears to interfere with terminal glycosylation of the angiotensin-converting enzyme 2 (ACE-2) cellular receptor, which is engaged by the virus for extracellular binding. This step may have a negative effect on the ability of the virus to gain entry into the host cell and, therefore, to initiate its replication cycle. Thus, infection may be deterred at clinically admissible concentrations [48]. Chloroquine also displays an immunomodulatory activity by suppressing the production and release of tumour necrosis factor alpha and interleukin 6 [49].

Furthermore, chloroquine was demonstrated to have strong antiviral activity against HCoV-OC43 *in vitro*. The anticoronaviral properties of chloroquine were also tested against HCoV-OC43 infection in newborn mice *in vivo*. Treatment with daily doses of chloroquine were found to have a long-lasting protective effect against lethal coronavirus OC43 infection in the newborn mice [1].

These favourable results suggest that chloroquine may be considered for use at antimalarial doses in the prevention of infections caused by coronaviruses, particularly SARS-CoV-2, which utilizes ACE-2 receptors in order to gain entry into host cells like its close relative, SARS-CoV-1.

6.4 Glycyrrhizic acid

Licorice root has been a commonly used ingredient in both Ayurvedic and traditional Chinese medicine for centuries, particularly in cough and cold remedies. Twenty triterpenoids and nearly three hundred flavonoids have been isolated from this herb. Scientific studies have shown that these metabolites possess many pharmacological activities including antiviral, antimicrobial, anti-inflammatory, and anti-tumour properties. However, glycyrrhizic acid or glycyrrhizin (GL), 18β-glycyrrhetinic acid (GA), liquiritigenin (LTG), licochalcone A (LCA), licochalcone E (LCE) and glabridin (GLD) are the main active components which possess antiviral and antimicrobial activities [50].

It has been known for some time that glycyrrhizic acid extracted from licorice (*Glycyrrhiza glabra*) root is active against viruses. This chemical is able to disrupt the growth and cytopathology of several unrelated DNA and RNA viruses without harming the host cell or its ability to replicate. Glycyrrhizic acid has also been demonstrated to inactivate herpes simplex virus particles irreversibly [51].

In a more recent study, the anti-SARSCoV activity of 15 glycyrrhizic acid derivatives was tested. Glycyrrhizin was shown to inhibit SARS-CoV-1 replication *in vitro* [52]. GL has also been reported to act by inhibiting viral gene expression and replication, reducing adhesion force and stress, and reducing High mobility group box 1 protein (HMGB1) binding to DNA. In addition, GL can enhance host cell activity by blocking the degradation of IκB, activating T lymphocyte proliferation and/or suppressing host cell apoptosis [50]. Thus, the potential for this licorice root component (GL) against SARS-CoV-2 infection is plausible.

6.5 Genistein & soy isoflavones

Isoflavones and their related flavonoid compounds, particularly genistein, exert antiviral properties against a wide range of DNA and RNA viruses *in vitro* and *in vivo* [53]. The biological properties of the flavonoids are well studied, but the mechanisms of action underlying their antiviral properties are not fully understood. Isoflavones appear to have a combination of negative effects on viruses including affecting virus binding, entry, replication, viral protein translation and formation of certain viral envelope glycoprotein complexes. A variety of host cell signalling processes can also be affected by isoflavones including induction of gene transcription factors and secretion of cytokines. All these effects are dependent on dose, frequency of administration, and different combinations of isoflavones employed in bioassays *in vitro*. Genistein may be able to mimic the action of 17-beta-estradiol [E2] due to its similar structure or to act as an E2 antagonist and its activity as a broad-spectrum tyrosine kinase inhibitor may contribute to its ability to influence estrogen receptor-independent mechanisms [54]. Despite their unique effect on immune function and anti-inflammatory activity, there is still a lack of data confirming the antiviral efficacy of such soy isoflavones *in vivo* against coronaviruses and other viruses thereby forming a worthwhile subject for biochemical study.

7. Summary

At least seven human coronaviruses have been described to date including SARS-CoV-2, which is closely related to and resembles SARS-CoV-1 in many respects. Both viruses bind to ACE-2 receptors on human cells. ACE2 is a membrane-associated aminopeptidase that converts angiotensin II to angiotensin 1–7 and plays a general role in the cleavage of peptides. Expression of ACE2 in human tissues correlates with known sites of SARS-CoV-1 infection including lungs (particularly airway epithelia), heart, kidneys, small intestine, testes, and vascular endothelia. These same tissues overlap with known sites of SARS-CoV-2 infection in humans.

A cluster of unexplained pneumonia cases were first reported in Wuhan, China and, a few days later, the cause of this pneumonia was identified as a new member of the coronavirus family. SARS-CoV-2 infection appears to be associated with a puzzling array of symptoms and complications. The major symptoms noted in China were fever (88%) and dry cough (67.7%), while rhinorrhea (4.9%) and gastrointestinal symptoms (diarrhea 4–14%) were less common. A majority of patients

(81%) had only mild symptoms (no pneumonia or mild pneumonia). Among patients with more pronounced symptoms, 14% experienced severe symptoms while 5% were critically ill with respiratory failure, septic shock, or multiorgan dysfunction or failure.

New data suggests that SARS-CoV-2 may spread via the bloodstream to infect other organs. In addition to the lungs, other target organs can include the heart, liver, brain, kidneys, and spleen. It also appears that COVID-19 may predispose patients to venous thromboembolism in several different ways including via endothelial dysfunction and promotion of a procoagulatory state. In fact, it was found that a significant percent of critically ill patients experienced myocardial injury from the infection and it has been concluded that COVID-19 may predispose to both venous and arterial thromboembolism due to excessive inflammation, hypoxia, immobilization, and diffuse intravascular coagulation. The COVID-19 pandemic is associated with various neurological symptoms and complications including anosmia, hypogeusia, seizures, and stroke, as well. COVID-19 complications in the brain can include delirium, inflammation, and encephalitis. Despite initial reports that children were unaffected by the novel coronavirus, it has emerged that pediatric patients are susceptible to a COVID-19-associated multisystem inflammatory syndrome that can cause serious inflammation of the blood vessels.

Several internal risk factors have been identified for SARS-CoV-2 infection. The main ones include age (older adults are more vulnerable to serious infection by the virus), gender (the virus is significantly more deadly in men than in women), obesity, heart disease, diabetes, cancer status, and smoking. However, there is no convincing evidence to date that any particular ethnic group displays a stronger genetic susceptibility to the virus (although, there may be a possible link to an inherited Neanderthal gene locus). Nevertheless, specific genetic variants such as those for the gene that encodes a protein that interacts with the ACE-2 receptor may be involved in determining individual patient responses to the disease. Simultaneously, external risk factors like environmental pollution, social conditions such as crowding, and frequency of exposure to infected persons also seem to play an important role.

Reports of re-infection in recovered humans is a surprising aspect of this virus. Recently, a team from the University of Hong Kong reported the first case of re-infection of COVID-19 within a period of approximately four and a half months. Genomic analyses confirmed that the patient had re-infection instead of persistent viral shedding from first infection. Moreover, there was a difference of 24 nucleotides between both viruses that infected the patient suggesting two different viral strains were involved [55].

Even though the virus is associated with positive COVID-19/COVID-19 antibody and high C-reactive protein test results, antibody levels may decline soon after infection. Consequently, it is quite possible that a lasting resistance to the virus will not be achievable. In the event that long-term immunity cannot be induced to the novel coronavirus by a vaccine, an annual, bi-annual, or even tri-annual inoculation may be required (current data suggests that antibodies begin to decrease or disappear three months after infection). This means that other modes of protection and prevention like supplementation may be more relevant in this case. Some candidates include Vitamin D, zinc, chloroquine/quinine, glycyrrhizic acid, and genistein due to anti-viral properties such as the ability to inhibit replication and reproduction of coronaviruses.

Scientists have concluded that drastic social distancing, quick detection and isolation of infected individuals and travel restrictions were the most effective steps for containment of COVID-19 in China. Genome sequencing has also helped to track and control COVID-19 infections quickly. However, if people do not continue to be careful, certain places may become vulnerable to further rounds of this disease. WHO recently reported that coronavirus infections among younger populations were skyrocketing. The proportion of cases in teens and young adults increased six-fold, while the proportion in young children and babies increased seven-fold by August. This may be attributable in part to the resurgence of large parties and social gatherings attended by young people following the relaxation of restrictions during the summer. Therefore, it seems very likely that the denouement of the COVID-19 story will be largely dictated by our social habits and ability to adapt to a new set of societal norms and conditions. This will include wearing face masks in public places, possibly, with a thin zinc coating along with a special zinc oxide crème formulation applied to the skin underneath [56].

Author details

Chanda Siddoo-Atwal[1,2,3]

1 University College London, London, United Kingdom

2 Simon Fraser University, Burnaby, Canada

3 President and Primary Biochemist of Moondust Cosmetics Ltd, West Vancouver, Canada

*Address all correspondence to: moondustcosmetics@gmail.com

References

[1] Keyaerts E, Li S, Vijgen L, Rysman E, Verbeeck J, Van Ranst M, Maes P. Antiviral Activity of Chloroquine against Human Coronavirus OC43 Infection in Newborn Mice. Antimicrobial Agents and Chemotherapy. 2009, DOI: 10.1128/AAC.01509-08

[2] Wang H, Li X, Li T, Zhang S, Wang L, Wu X, Liu J. The Genetic Sequence, Origin, and Diagnosis of SARS-CoV-2. European Journal of Clinical Microbiology and Infectious Diseases. 2020, Apr 24: 1-7. DOI: 10.1007/s10096-020-03899-4

[3] Clerkin KJ, Fried JA, Raikhelkar J, Sayer G, Griffin JM, Masoumi A, Jain SS, Burkhoff D, Kumaraiah D, Rabbani L, Schwartz A, Uriel N. COVID-19 and Cardiovascular Disease. Circulation. 2020; 141: 1648-1655. DOI.org/10.1161/CIRCULATIONAHA.120.046941

[4] Zhang Q, Cui J, Huang X, Zheng H, Huang J, Fang L, Li K, Zhang J. The Life cycle of SARS Coronavirus in Vero E6 Cells. Journal of Medical Virology. 2004, 73: 332-337. DOI: 10.1002/jmv.20095

[5] Hoffmann M, Kleine-Weber H, Schroeder S, et al. SARS-CoV-2 Cell Entry Depends on ACE2 andTMPRSS2 and Is Blocked by a Clinically Proven Protease Inhibitor. Cell. 2020, 181: 271-280. DOI.org/10.1016/j.cell.2020.02.052

[6] He F, Deng Y, Li W. Coronavirus disease 2019: What we know? Journal of Medical Virology. 2020, 92(7): 719-725. DOI.org/10.1002/jmv.25766

[7] Jia HP, Look DC, Shi L, Hickey M, Pewe L, Netland J, Farzan M, Wohlford-Lenane C, Perlman S, McCray PB Jr. ACE2 Receptor Expression and Severe Acute Respiratory Syndrome Coronavirus Infection Depend on Differentiation of Human Airway Epithelia. Journal of Virology. 2005, 79(23): 14614-14621. DOI: 10.1128/JVI.79.23.14614-14621.2005

[8] Deng W, Bao L, Gao H, Xiao C, Liu J, Xue J, Lv Q, Liu J, Yu P, Xu Y, Qi F, Qu Y, Li F, Xiang Z, Yu H, Gong S, Liu M, Wang G, Wang S, Song Z, Zhao W, Han Y, Zhao L, Liu X, Wei Q, Qin C. Reinfection could not occur in SARS-CoV-2 infected rhesus macaques. bioRxiv (Cold Spring Harbor Laboratory). March 2020. DOI .org/10.1101/2020.03.13.990226

[9] World Health Organization; www.who.int coronavirus (Q&A) – Older People and COVID-19

[10] Richardson S, Hirsch JS, Narasimhan M, et al. Presenting Characteristics, Comorbidities, and Outcomes Among 5700 Patients Hospitalized With COVID-19 in the New York City Area. Journal of the American Medical Association. 2020, 323(20): 2052-2059. DOI:10.1001/jama.2020.6775

[11] Zheng Z, Peng F, Xu B, et al. Risk factors of critical & mortal COVID-19 cases: A systematic literature review and meta-analysis. Journal of Infection. 2020, 81(2): e16-e25. DOI.org/10.1016/j.jinf.2020.04.021

[12] Wichmann D, Sperhake J-P, Lutgehetmann M, et al. Autopsy Findings and Venous Thromboembolism in Patients with COVID-19; A Prospective Cohort Study. Annals of Internal Medicine. 2020, May 6: M20-2003. DOI:10.7326/M20-2003

[13] Dai M, Liu D, Liu M, et al. Patients with Cancer Appear More Vulnerable to SARS-CoV-2: A Multicenter Study during the COVID-19 Outbreak. Cancer Discovery, 2020, 10(6): 783-701. DOI: 10.1158/2159-8290.CD-20-0422

[14] Ng J, Bakrania K, Falkous C, Russell, R. COVID-19 Mortality Rates by Age and Gender: Why Is the Disease Killing More Men than Women? RGA. 2020

[15] Verma S, Saksena S, Sadri-Ardekani H. ACE2 receptor expression in testes: implications in coronavirus disease 2019 pathogenesis. Biology of Reproduction. 2020, 103(3): 449-451. DOI.org/10.1093/biolre/ioaa080

[16] Berlin I, Thomas D, Le Faou A-L, Cornuz J. COVID-19 and Smoking. Nicotine and Tobacco Research. 2020, 22(9): 1650-1652. DOI.org/10.1093/ntr/ntaa059

[17] Bhala N, Curry G, Martineau AR, Agyemang C, Bhopal R. Sharpening the global focus on ethnicity and race in the time of COVID-19. The Lancet. 2020, 395 (10238): 1673-1676. doi.org/10.1016/S0140-6736(20)31102-8

[18] Siddoo-Atwal C. A Case-Study of Apoptotic Sunburn with Sunscreen. Indian Journal of Ecology. 2011, 38(2): 300-301

[19] Siddoo-Atwal C. A Case-Study of Two Sunscreens that May Prevent Apoptotic Sunburn. Indian Journal of Ecology. 2012, 39(1): 131-134

[20] Siddoo-Atwal C (2019A) Chapter 4 - Genes That Can Cause Cancer in "Gene Expression Profiling in Cancer" INTECH (edited by Vlachakis D). DOI: 10.5772/intechopen.83549

[21] Siddoo-Atwal, C. AT, Apoptosis, and Cancer: A Viewpoint. Indian Journal of Ecology, 2009. 36(2): 103-110

[22] Price M. Africans carry a surprising amount of Neanderthal DNA. Science. January 30, 2020. doi:10.1126/science.abb0984

[23] Zeberg, H., Pääbo, S. The major genetic risk factor for severe COVID-19 is inherited from Neanderthals. Nature (2020). https://doi.org/10.1038/s41586-020-2818-3

[24] Grant WB, Lahore H, McDonnell SL, Baggerly CA, French CB, Aliano JL, Bhattoa HP. Evidence that Vitamin D Supplementation Could Reduce Risk of Influenza and COVID-19 Infections and Deaths. 2020, Nutrients 12(4): 988. DOI.org/10.3390/nu12040988

[25] Setti L, Passarini F, De Gennaro G, Barbieri P, Perone MG, Piazzalunga A, Borelli M, Palmisani J, Di Gilio A, Piscitelli P, Miami A. The Potential Role of Particulate Matter in the Spreading of COVID-19 in Northern Italy: First Evidence-Based Research Hypotheses. 2020A, mediRxiv (Cold Spring Harbor Laboratory). DOI.org/10.1101/2020.04.I I. 20061713

[26] Setti L et al (as above). SARS-CoV-2 RNA Found on Particulate Matter of Bergamo in Northern Italy: First Preliminary Evidence. 2020B, medRxiv (Cold Spring Harbor Laboratory). DOI.org/10.1101/2020.04.15.20065995

[27] Klok FA, Kruip MJHA, Van Der Meer NJM, Arbous MS, Gommers DAMPJ, Kant KM, Kaptein FHJ, Van Passsen J, Stals MAM, Huisman MV, Endeman H. Incidence of thrombotic complications in critically ill ICU patients with COVID-19. 2020, Thrombosis Research 191 (2020): 145-147. DOI.org/10.1016/j.thromres.2020.04.013

[28] Hess DC, Eldeshan W, Rutkowski E. COVID-19-Related Stroke. Translational Stroke Research. 2020, 11(11): 1-4. DOI: 10.1007/s12975-020-00818-9

[29] Paterson RW, Brown RL, Benjamin L, et al. The emerging spectrum of COVID-19 neurology: clinical, radiological and laboratory findings. Brain (A Journal of Neurology), 2020. DOI.org/10.1093/brain, (2020)

[30] Brann DH, Tsukahara T, Weinreb C, et al. Non-neural expression of SARS-CoV-2 entry genes in the olfactory system suggests mechanisms underlying COVID-19-associated anosmia. Science Advances. July 24, 2020 as DOI:10.1126/sciadv.abc5801

[31] Mallapaty S. Mounting clues suggest the coronavirus may trigger diabetes. 2020, Nature 583(7814): 16-17. DOI:10.1038/d41586-020-01891-8

[32] Godfred-Cato S, Bryant B, Leung J et al. 2020, COVID-19-Associated Multisystem Inflammatory Syndrome in Children - United States March-July 2020. Morbidity and Mortality Weekly Report 69(32): 1074-1080. DOI:10.15585/mmwr.mm6932e2

[33] British Society for Immunology - BSI open letter to Government on SARS-CoV-2 outbreak response – 14 March 2020

[34] Bridle BW & Sharif S. Fast COVID-19 timelines are unrealistic and put the integrity of scientists at risk. The Conversation. June 15, 2020

[35] Zhu Q & Berzofsky JA. Oral Vaccines; Directed safe passage to the front line of defense. 2013, Gut Microbes 4(3): 246-252. DOI: 10.4161/gmic.24197.

[36] Martineau AR, Jolliffe DA, Hooper RL, et al. Vitamin D supplementation to prevent acute respiratory tract infections: systematic review and meta-analysis of individual participant data. 2017, British Medical Journal 356. DOI.org/10.1136/bmj.i6583

[37] Ilie PC, Stefanescu S, Smith L. The role of Vitamin D in the prevention of coronavirus disease 2019 infection and mortality. 2020, Aging Clinical and Experimental Research 32: 1195-1198. DOI.org/10.1007/s40520-020-01570-8

[38] Velthuis AJW, van den Worm SHE, Sims AC, Baric RS, Snider EJ, van Hemert MJ. Zn^{2+} Inhibits Coronavirus and Arterivirus RNA Polymerase Activity *In Vitro* and Zinc Ionophores Block the Replication of These Viruses in Cell Culture. PLOS Pathogens; Published: November 4, 2010 https://doi.org/10.1371/journal.ppat.1001176

[39] Gao Y, Yan L, Huang Y, Liu F-J, et al (2020) Structure of the RNA-dependent RNA Polymerase from COVID-19 Virus. Science 368(6492):779-782. doi: 10.1126/science.abb7498

[40] Ruttkay-Nedecky B, Nejdl L, Gumulrc J, Zitka O, Masarik M, Eckschlager T, Stiborova M, Adam V, Kizek R (2013) The Role of Metallothionein in Oxidative Stress. International Journal of Molecular Sciences 14(3): 6044-6066

[41] Siddoo-Atwal C (2019B) Chapter 17 – Biological Effects of Uranium and Its Decay Products on Soil Microbes, Plants, and Human; in "Plant Microbe Interface" Springer (edited by Varma A, Tripathi S & Prasad R)

[42] Lucia CMD, Santos LLM, da Cruz Rodrigues KC, da Cruz Rodrigues VC, Martino HSD, Sant'Ana HMP (2014) Bioavailability of Zinc in Wistar Rats Fed with Rice Fortified with Zinc Oxide. Nutrients 6(6): 2279-2289

[43] Wegmuller R, Tay F, Zeder C, Brnic M, Hurrell RF (2014) Zinc Absorption by Young Adults from Supplemental Zinc Citrate Is Comparable with That from Zinc Gluconate and Higher than from Zinc Oxide. The Journal of Nutrition 144(2): 132-136

[44] Agren MS. Percutaneous absorption of zinc from zinc oxide applied topically to intact skin in man. Dermatologica 180: 36-39, 1990; https://doi.org/10.1159/000247982

[45] Siddoo-Atwal C. A Case-Study of Uranium and Heavy Metal Detoxification. Indian Journal of Ecology. 2019C, 46(2): 427-430

[46] Abolghasemi E. Mossa-Kazemi SH, Davoudi M, Reisi A, Satvat MT. Comparative study of chloroquine and quinine on malaria rodents and their effects on the mouse testis. Asian Pacific Journal of Tropical Biomedicine. 2012, 2(4): 311-314. DOI: 10.1016/S2221-1691(12)60030-6

[47] Keyaerts E, Vijgen L, MAEs P, Neyts J, Van Ranst M. In vitro inhibition of severe acute respiratory syndrome coronavirus by chloroquine. 2004, Biochemical and Biophysical Research Communications 323(1): 264-268. DOI: 10.1016/j.bbrc. 2004.08.085

[48] Vincent MJ, Bergeron E, Benjannet S, Erickson, BR, Rollin PE, Ksiazek TG, Seida DG, Nichol ST. Chloroquine is a potent inhibitor of coronavirus infection and spread. 2005, Virology Journal 2: 69. DOI: 10.1186/1743-422X-2-69

[49] Savarino A, Boelaert JR, Cassoni A, Majori G, Cauda R. Effects of chloroquine on viral infections: an old drug against today's diseases. 2003, The Lancet (Infectious Diseases) 3(11): 722-727

[50] Wang L, Yang R, Yuan B, Liu Y, Liu C. The antiviral and antimicrobial activities of licorice, a widely-used Chinese herb. 2015, Acta Pharmaceutica Simica B 5(4): 310-315. DOI.org/10.1016/j.apsb.2015.05.005

[51] Pompei R, Flore O, Marccialis MA, Pani A, Loddo B. Glycyrrhizic acid inhibits virus growth and inactivates virus particles. 1979, Nature 281(5733): 689-690. DOI.org/10.1038/281689a0

[52] Hoever G, Baltina L, Michaelis M, Kondratenko R, Baltina L, Tolstikov GA, Doerr HW, Cinatl J Jr. Antiviral activity of glycyrrhizic acid derivatives against SARS-coronavirus. 2005, Journal of Medicinal Chemistry 48(4): 1256-1259. DOI.10.1021/jm0493008

[53] Andres A, Donovan SM, Kuhlenschmidt MS. Soy isoflavones and virus infections. 2009, Journal of Nutritional Biochemistry 20(8): 563-569. DOI:10.1016/jnutbio.2009.04.004

[54] Sakai T & Kogiso M. Soy Isoflavones and Immunity. 2008, The Journal of Medical Investigation 55(3-4): 167-173. DOI: 10.2152/jmi.55.167

[55] Garcia de Jesus E. A Man in Hong Kong is the first confirmed case of coronavirus reinfection. ScienceNews. August 24, 2020

[56] Siddoo-Atwal C. A Proposed Weapon in the Anti-COVID-19 Arsenal: Percutaneous Zinc Absorption. 2020, Clinical Case Reports 10 (9): 1380. DOI: 10.37421/jccr.2020.10.1380

Chapter 3

How Can We Be Ahead of COVID-19 Curve? A Hybrid Knowledge-Based and Modified Regression Analysis Approach for COVID-19 Tracking in USA

Rafaat Hussein

Abstract

Since its appearance in 2019, the COVID-19 virus deluged the world with unprecedented data in short time. Despite the countless worldwide pertinent studies and advanced technologies, the spread has been neither contained nor defeated. In fact, there is a recent record surge in the number of confirmed new cases. The rational question is thus: why has it taken so long to date to forecast the trajectory of the spread? To this end, this chapter presents a new predictive Knowledge-based (KB) toolkit named CORVITT (Corona Virus Tracking Toolkit) and a modified linear regression model. This logical step assists the officials, organizations, and users to forecast the spread trajectory and accordingly make proactive rather than retroactive intervention decisions. This hybrid approach uses the confirmed new cases and demographic data, implemented. CORVITT is not an epidemiological model, in the sense that it does not model disease transmission, nor does it use underlying epidemiological parameters or data including the reproductive rate, disease methods, real time polymerase chain reaction cycle threshold, the virus structure and pathogenesis, etc. The chapter is a seed in an in-progress study that will broaden its scope by including additional parameters.

Keywords: COVID-19, coronavirus, epidemic, modelling, outbreak, pandemic

1. Introduction

The unfolding COVID-19 has turned the world upside down [1], and this unprecedented trend is set to be the worst pandemic of a generation in terms of the increasing number of infected people. In its report on April 7, 2020, the US Centers for Disease Control and Prevention (CDC) [2] indicated that the COVID-19 poses a severe threat to public health. In its report, the CDC indicated that the "complete clinical picture with regard to COVID-19 is not fully known." To deal with this blurred picture, the World Health Organization (WHO) has compiled an overwhelmingly pertinent database [3]. The CDC provides a daily report that includes new data reported to the CDC by 55 USA jurisdictions [4]. Many other organizations have also provided similar resources and statistics including the Chinese

Medical Association Publishing House [5] and the European Centre for Disease Prevention and Control [6]. All the available approaches suggest that the number of new COVID-19 cases plays a key role in mapping its trajectory [7] worldwide.

COVID-19 is an evolving epidemic, and its up-and-down spread (trend or pattern commonly referred to as "curve") is a sign of its elusiveness. As of today (July 25, 2020), the COVID-19 is striking back with record-setting blows. In general, the COVID-19 issue relates to various facets such as public health and social as well as culture characteristics, and the world seems lacking sound methodologies on how to address this problem. Using predictive tracking or forecasting quantitative measures can assist the authorities, officials, organizations, and users to be proactive rather than reactive, and thus better prepared to mitigate potential adversaries.

2. The model

The literature seems to suggest that using the number of new cases and the level of social distancing are the key variables to analyze the COVID-19 in various ways. In what follows, we provide a background information about the four main COVID-19 modeling techniques: system dynamics, agent-based modeling, discrete event simulation, and hybrid simulation [8]. System dynamics uses differential equations to model resources, knowledge, people, and money, and the flows between these parameters explains the simulation behavior. The agent-based techniques are stochastic, enabling the variability of human behavior to be incorporated to help understand the likely effectiveness of proposed protective measures. The discrete event technique is also stochastic and models operations over time where entities flow through a number of activities. The hybrid simulation combines two or more techniques and is used for complex behavior. These techniques focus mainly on the unfolding phases of disease transmission such as quarantine, lock down, testing, and health care services. Some of these approaches have been rooted in the literature since 1777, and are complex, and cumbersome to implement. Without adequate specialists in advanced and complex mathematical theories and/or computers, the logical question is thus: how could the proper personnel ascertain the COVID-19 spread in order to make proactive intervention decisions; e.g. to prepare hospitals and intensive care units, to mitigate the adverse impacts of what may happen in the near future? In search for accurate answer and based on the popular utilization of COVID-19 relationship between the number of cases [9] and population per land area, the idea of a new index was conceptualized in this study. It represents the number of reported confirmed new cases per population in the specific region the data was recorded. This new concept harnesses the number of cases and the regional crowdedness of people, which varies in the US from single digit to multi-thousand [2]. The index increases with more cases and with more dense populations (assumed shorter social distancing).

In this study, a combined linear regression analysis and data-fitting model is used. To deal with data fluctuation, this model adopted the hypothesis that was successfully used in other published studies of a short time span of one month maximum for forecasting, [10–13]. That hypothesis is logical and rational because the world knows that the virus spread in unpredictable; thus, longer time spans may encompass inaccurate data. The data is obtained from the New York Times Journal database [14]. The journal publishes the daily cases of COVID-19 by state and county in the US. The data from eleven states was used: New York State (NYS), Florida (FL), California (CA), Colorado (CO), Illinois (IL), Texas (Tx), Louisiana (LA), Washington (WA), Georgia (GA), New Jersey (NJ), and Michigan (MI).

We first constructed the slope for the confirmed cases and population in each of the states from March 27 to May 11, 2020, and then used polynomials fitting and linear regression analysis for forecasting. Linear regression is a direct way to deal with the connections between variables.

3. New knowledge-based toolkit

To accurately and proactively capture the big picture of COVID-19 spread, this study transfers the expertise of problem-solving from humans into a KB toolkit that takes in the same data, and yields the same conclusion but faster. This new KB-statistic hybrid approach effectively assists humans in dealing with COVID-19 massive daily data in addition to save time which is an essential requirement in dealing with the virus illusiveness. The study introduces for the first time in this field, to our knowledge, a novel KB toolkit to visualize the data and make it easier to understand and use without either mathematical or computer expertise. The CORVITT is a promising incubator for COVID-19 future forecasting platforms. Its VBA-based architecture blueprint emerges from an open-end modular adaptable structure encompassing a graphical-interface client allowing the users to easily operate it. This KB technology has been proven in other applications and thus applied in this study for COVID-19 [15–17]. To the author's knowledge, the concept of CORVITT has not been attempted to date for COVID-19. **Figure 1** shows the dashboard of CORVITT. The user could simply click the button that represents the state/province of interest, and the dashboard will display the microdata or the relative comparison of all states. **Figure 2** shows the data used in **Figure 1**. Although the amount of collected data is massive, the use of the dashboard is intuitive and user

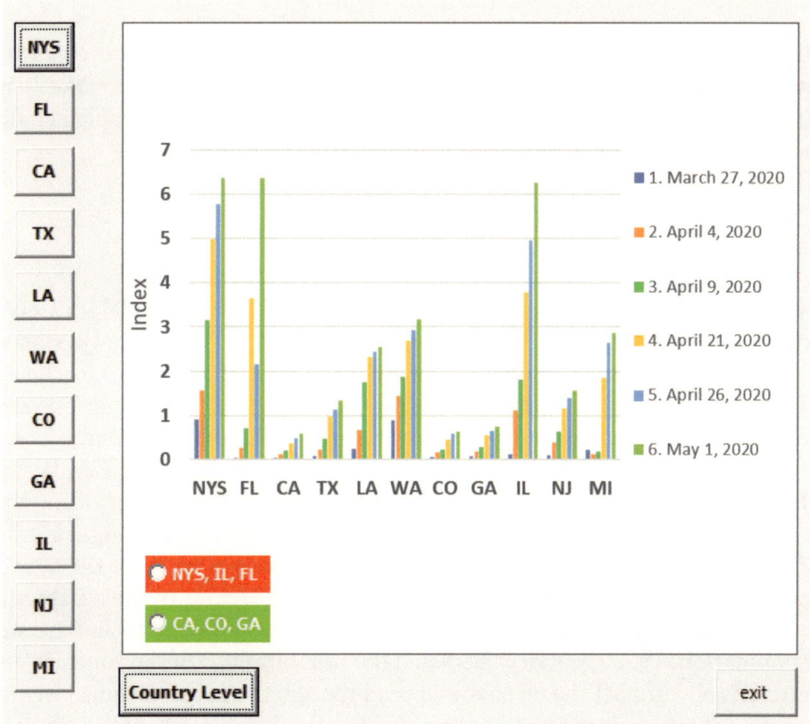

Figure 1.
Dashboard of the CORVITT presented in this chapter.

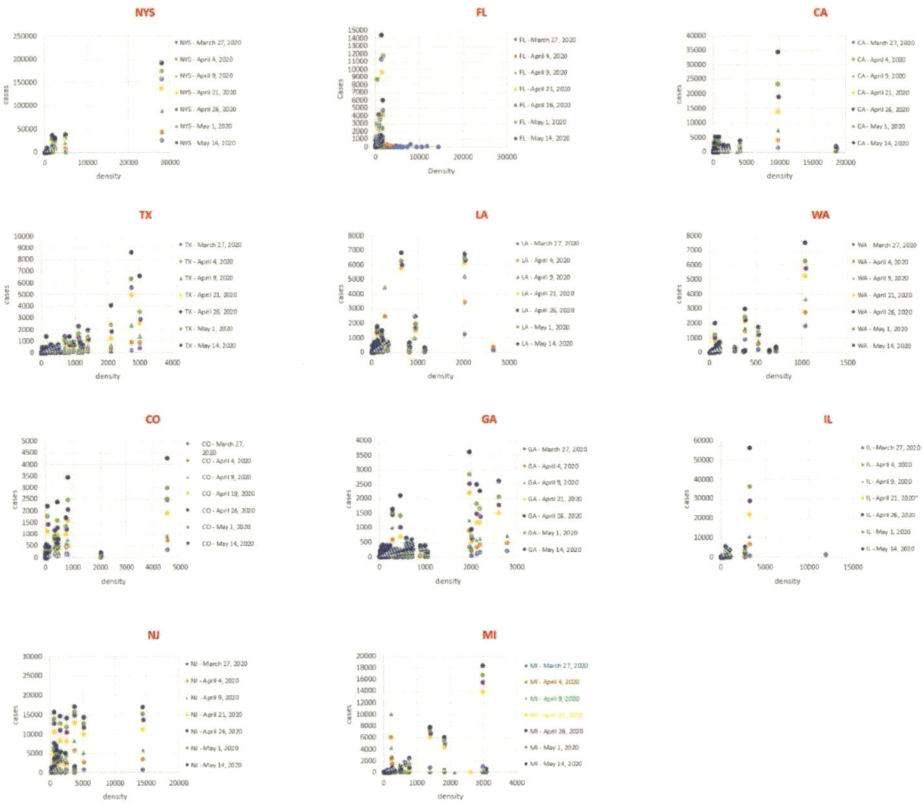

Figure 2.
The macro-data used in this study.

friendly. Again, there is no need for medical, mathematical, or computer skills to use the dashboard and benefit from its applications. This is one of the takeaways of bringing an artificial brain to help human brains in dealing with complex challenge at hand such as the COVID-19.

4. Results and discussion

In what follows, we examined the feasibility of the ascribed model for COVID-19 in two ways: firstly, by analyzing its forecasted outcomes in eleven US states, and secondly by comparing the forecasted results with actual onsite data. Firstly, a database was created at micro-level or counties, for the first time to our knowledge, for the new cases and population per land area on COVID-19 from March 27 to May 1, 2020 in the eleven US states: NYS, FL, CA, TX, LA, WA, CO, GA, IL, NJ, and MI. These sates had a steady high number of confirmed cases according to the New York Times Journal. **Table 1** shows some of the collected data. **Figure 2** shows CORVITT gauging of the virus county-wise distributions in terms of the new index and population. **Figure 2** shows that in NYS, FL, CA, CO, and IL the inhabitants are infected in areas with large social distances because most of the data is concentrated at low population, i.e. large spaces between the inhabitants. On the contrary, in TX, LA, WA, GA, NJ, and MI the virus was spread though the social distance was small because the data spreads over a wide range of population, i.e. large space between the inhabitants. Unlike the common general approach that was used for all US states

Regression statistics						
Multiple R				0.992092		
R Square				0.984246		
Adjusted R Square				0.976369		
Standard error				0.372065		
Observations				4		
ANOVA						
	df	SS	MS	F	Significance F	
Regression	1	17.29721	17.29721	124.9503	0.007908	
Residual	2	0.276865	0.138433			
Total	3	17.57407				
	Coefficients	**Standard error**	**t Stat**	**P-value**	**Lower 95%**	**Upper 95.0%**
Intercept	-7476.32	669.1315	-11.1732	0.007915	-10355.4	-4597.28
date	0.170252	0.015231	11.17812	0.007908	0.104719	0.235785

Table 1.
A sample output of the modified regression.

at all times since 2019, which is common in the news media from Tabloid to New York Times journals, this discovery unveiled new facets. **Figure 1(d)** shows that on March 27, the indexes were 0.93 and 0.10 in NYS and NJ (large distance), respectively, 0.08 and 0.07 for TX and GA (small social distance), respectively, though the spread of data appeared similar on the dashboard in each group. In addition, NYS, LA, WA, IL, and MI have high indexes by comparison to FL, CA, TX, CO, GA, and NJ For example, on April 21, 2020, the indexes were 5.0 and 0.40 for NYS and CA although the closeness of data in both states was similar. Furthermore, the increase of the index within each state was nonuniform. For example, the index increased from 2.8 to 5 between April 9 and 21 in NYS, and from 0.05 to 1.0 in TX over the same period. From a different angel of view, **Figure 1(a)** shows that the rate of spreading of the virus differ from one state to another. For example, the spreading rate in IL is very high compared to that in CA. This indicates that the virus can affect more people in IL (57,920 sq. mi and 13 million people) than NJ (8730 sq. mi and 9 million people). Taken as a whole, CORVITT-outcomes suggest that NYS is a good region for the virus to spread whereas CA is not as good from March 27 to May 1, 2020. Such information allows the authorities to prioritize the resources giving NYS the highest priority. Secondly, **Figure 3a** and **b** compares the forecasted and actual on-site data and shows a close agreement in different US states. **Figure 3a** shows that whereas the new cases in NYS has reached the peak in the first week of May, the pandemic was worsening in other states such as Illinois, but in California, Georgia, and Colorado reached a plateau. **Figure 3c** shows the severity of the pandemic as indicated by the skewness; positive (to the right) for LA and negative for NYS, NJ, and MI of the growth and deterioration of the distributions [18]. The positive skewness means longer deterioration (decline in cases) time. The shallow deterioration rate at the trailing end of the curve in **Figure 1(b)** is a sign of a plateau. **Figure 1** describes the peak, weakness, and steadiness statuses by which the virus trajectory disperses through different stages in various regions. This new discovery is useful to understand the building up and collapse of the virus impacts thus make proactive preparations possible.

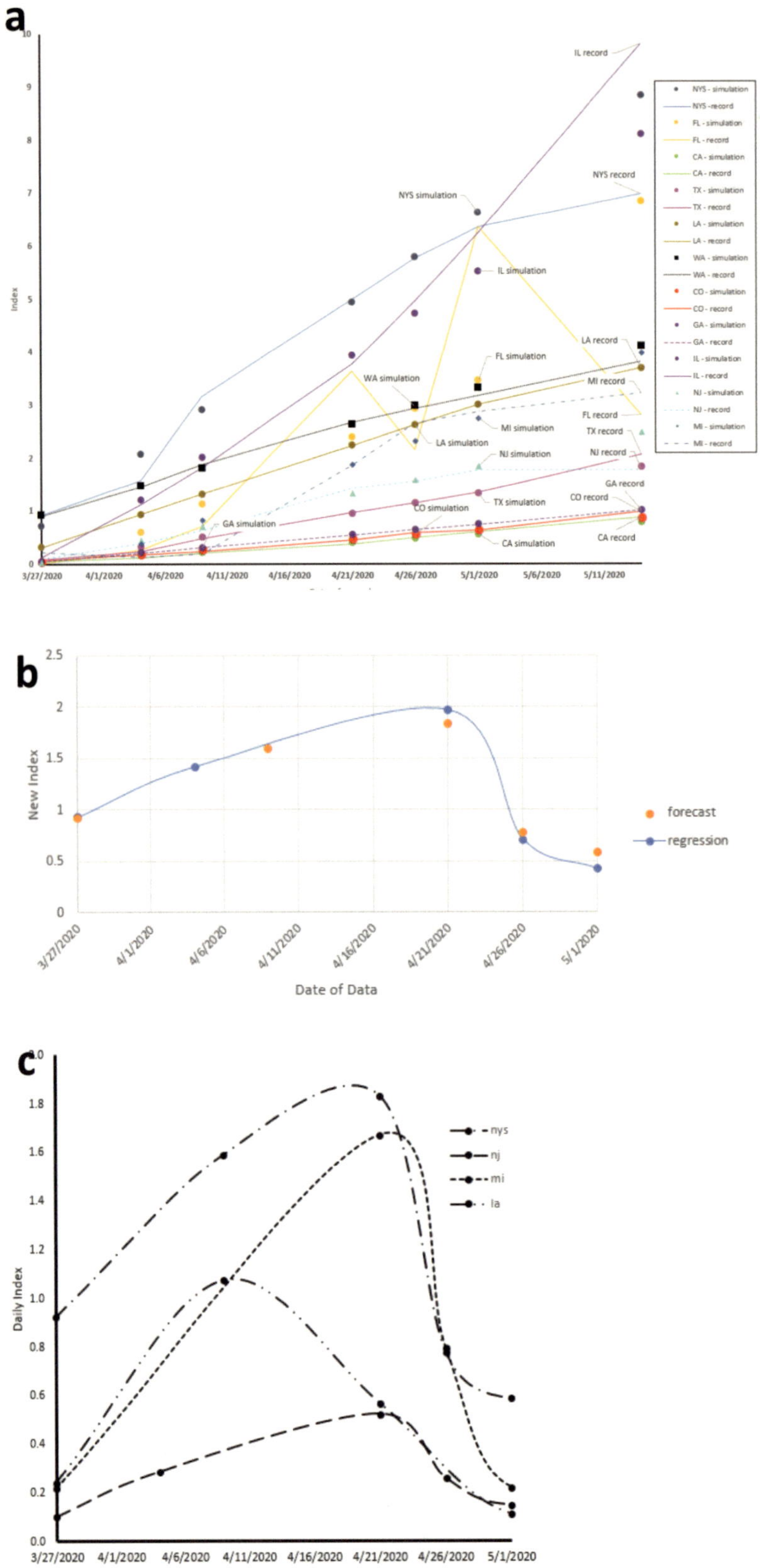

Figure 3.
(a): The trajectory of cases' dispersion over time in various US States. (b): A satisfactory agreement between the forecasted and actual data. (c). The growth and deterioration distributions of cases over time.

5. Conclusions

As the enormity of the COVID-19 threat has become clear, the characteristics of existing COVID-19 complex analytic methodologies and the all-encompassing approach place serious limitations on their usefulness for practical use. The computer technologies have reached what no one could imagined, and the KB systems have proven very beneficial in many fields. The rational question is: why has it taken so long for a logical approach to appear to practicalize the analytical complex simulations? To answer the question, this chapter introduces machine smartness to assist humans' intelligence to capture the big picture of the virus illusiveness thus take proactive rather than retroactive steps to mitigate safely its inevitable adverse effects. This seed study introduced a hybrid KB-regression analysis model for COVID-19 forecasting. It used data collected from eleven US states at macro-level level to foresee the short-term spread trajectory. The outputs unveiled new discoveries and shed light on various facets of the COVID-19 in each state. The accuracy of the hybrid approach was gauged by comparing forecasted and actual data and satisfactory agreements were found. It should be noted that this study is a step forward, but additional development is in progress for improvement preparations.

Author details

Rafaat Hussein
State University of New York, Syracuse NY, US

*Address all correspondence to: ezpsc@yahoo.com

References

[1] United nations Committee for the Coordination of Statistical Activities, How covid-19 is changing the world, 2020. https://unstats.un.org/unsd/ccsa/documents/covid19-report-ccsa.pdf.

[2] The Centers for Disease Control and Prevention [CDC], Situation Summary, 2020. https://www.cdc.gov/coronavirus/2019-ncov/cases-updates/summary.html.

[3] The World Health Organization [WHO], Global research on coronavirus disease [covid-19], 2020. https://www.who.int/emergencies/diseases/novel-coronavirus-2019/global-research-on-novel-coronavirus-2019-ncov.

[4] The Centers for Disease Control and Prevention [CDC], Cases in U.S., 2020. https://www.cdc.gov/coronavirus/2019-ncov/cases-updates/cases-in-us.html.

[5] The Chinese Medical Association Publishing House, COVID-19 Academic Research Platform, 2020. http://eng.med.wanfangdata.com.cn/Chinese_Medical_Association_Publishing_House.html.

[6] The European Centre for Disease Prevention and Control, Coronavirus disease, 2020. https://www.ecdc.europa.eu/en.

[7] STAT, The coronavirus is washing over the U.S. These factors will determine how bad it gets in each community, 2020. https://www.statnews.com/2020/04/01/coronavirus-how-bad-it-gets-different-communities/

[8] Currie, C. et al., How simulation modelling can help reduce the impact of covid-19, Journal of Simulation, Vol. 14, pp: 83-97, 2020. https://www.tandfonline.com/doi/pdf/10.1080/17477778.2020.1751570?needAccess=true

[9] Our World in Data, COVID-19 death rate vs. Population density, 2020. https://ourworldindata.org/grapher/covid-19-death-rate-vs-population-density

[10] Yousaf, M. et al., Statistical analysis of forecasting covid-19 for upcoming month in Pakistan, Elsevier Public Health Emergency Collection, 2020. DOI: 10.1016/j.chaos.2020.109926

[11] Sujath, R. et al., A machine learning forecasting model for covid-19 pandemic in India, Stochastic Environmental Research and Risk Assessment, 2020. https://doi.org/10.1007/s00477-020-01843-8.

[12] Yonar H. et al., Modeling and Forecasting for the number of cases of the covid-19 pandemic with the Curve Estimation Models, the Box-Jenkins and Exponential Smoothing Methods, Eurasian Journal of medicine and Oncology, pp: 160-165, 2020. DOI: 10.14744/ejmo.2020.28273

[13] Öztoprak,F. and Javed, A, Case Fatality Rate estimation of covid-19 for European Countries: Turkey's Current Scenario Amidst a Global Pandemic; Comparison of Outbreaks with European Countries, Eurasian Journal of medicine and Oncology, Vol. 4 No. 2, pp: 149-159, 2020. DOI: 10.14744/ejmo.2020.60998

[14] The New York Times Journal, The Coronavirus in the US, 2020. https://www.nytimes.com/interactive/2020/us/coronavirus-us-cases.html

[15] Hussein, R., Knowledge-based & expert system technologies for built & natural environments, 2020, https://aitechm.wixsite.com/8kbextech.

[16] Hussein, R., Artificial Intelligence for the Built and Natural Environments

and Climate, In: Advances in Artificial Intelligence, edited by Sergey Yurish, the International Frequency Sensor Association, 2019, Chapter 7.

[17] Hussein, R., Knowledge-based Tools for Monitoring and Management of the Engineered Infrastructure Construction Systems, In: Advances in Computers and Software Engineering, edited by Sergey Yurish, International Frequency Sensor Association, 2020, Chapter 5.

[18] U.S. Department of Health and Human Services, Principles of Epidemiology in Public Heath Practice, 2012. https://www.cdc.gov/csels/dsepd/ss1978/index.html

Chapter 4

Renin Angiotensin System, Gut-Lung Cross Talk and Microbiota. Lessons from SARS-CoV Infections

Andreia Matos, Alda Pereira da Silva, Joana Ferreira, Ana Carolina Santos, Maria Clara Bicho and Manuel Bicho

Abstract

The two antagonistic systems of angiotensin converting enzyme (ACE)-1 and ACE-2 are in the "eye of the hurricane" of severe acute respiratory syndrome coronavirus (SARS-CoV-2). The receptor of the SARS-CoV-2 is the same as ACE-2, which causes its under-expression after binding it, followed by the internalization of the complex virus-ACE-2. ACE-2 have multiple functions with specially relevance in cardiovascular diseases. Furthermore, the non-enzymatic role of ACE-2 gives rise to a Hartnup disease, a phenocopy involving microbiota. With this chapter, we intent to explore the key pathways involved in SARS-CoV-2 infection, from the host perspective, considering our hypothesis related to transporter of neutral amino acids, which includes tryptophan precursor of serotonin and kynurenine.

Keywords: severe acute respiratory syndrome coronavirus (SARS-CoV-2), renin-angiotensin system, tryptophan precursors, microbiome, genetic susceptibility

1. Background

Severe acute respiratory syndrome coronavirus 2 (SARS-CoV-2), initiated in Wuhan in China, revisited 17 years later an outbreak that started in 2002 in China caused by a virus very similar to SARS-CoV-2 [1]. Identification and sequencing of the virus responsible for COVID-19 determined that it was a novel coronavirus that shared 88% sequence identity with two bat-derived SARS-like CoV, suggesting it's origin in bats [2]. Additionally, it was shown that this coronavirus, which was termed 2019-nCoV or SARS-CoV-2, shared 79.5% sequence identity with SARS-CoV [2].

After inhalation of SARS-CoV-2, it invades nasal epithelial cells (superior respiratory tract) and type II pneumocytes through binding the SARS spike protein to angiotensin-converting enzyme 2 (ACE-2) receptors [3]. This complex is proteolytically processed by transmembrane protease serine 2 (TMPRSS2), leading to cleavage of ACE-2 and activation of the spike protein, thereby facilitating viral entry into the target cell. For SARS-CoV-2 entry into a host cell, its spike protein needs to be cleaved by cellular proteases at 2 sites, termed S protein priming by the serine protease TMPRSS2, then the viral and cellular membranes can fuse [4]. It has been suggested that cells in which both ACE-2 and TMPRSS2 are expressed are most susceptible to entry by coronaviruses from the SARS family, among which is the virus described to cause SARS and, also SARS-CoV-2 [4, 5].

In relation to the mechanism of infection, the infected cells trigger the host's immune response, and the inflammatory cascade is initiated by innate immune cells, being the host environment extremely important for internalization and multiplication of the virus [6]. Possible mechanisms of receptor and signaling mechanisms responsible for induction of inflammatory mediators, such as cytokines or chemokines, may be related to the release of danger signal molecules, like certain cytokines, or may be involve a different recognition pathway mediated by immune cells throughout known pattern recognition receptors, such as toll-like receptors (TLRs) [7].

The heterologous protection against infections through epigenetic, transcriptional, and functional reprogramming of innate immune cells may contribute to different susceptibility to severity of SARS-CoV-2 [7, 8]. Furthermore, the changes in metabolic and endocrine pathways associated with SARS-CoV-2 infection may untangle a more profound understanding of this disease and contribute to a more adequate response.

2. Host systemic reactions of SARS-CoV-2 infection

Although the SARS-CoV-2 infection is highly associated to respiratory infection, it is also true, that this infection reflects a systemic involvement with multiple symptoms, including fever, persistent dry cough, shortness of breath, chills, muscle pain, headache, loss of taste or smell, and gastrointestinal symptoms [9]. Interestingly, according to the clinical features of individuals affected with SAR-CoV-2, a significant proportion of patients initially present some atypical gastrointestinal symptoms such as diarrhea, nausea, and vomiting [10].

Coronaviruses are one of many pathogens known to cause postinfectious olfactory dysfunction, nasal epithelial cells and mainly goblet cells in a high expression patterns of the ACE-2 receptor, which is required for SARS-CoV-2 entry. Olfactory dysfunction and anosmia are highly implicated in SARS-CoV-2 infection. The inclusion of loss of smell or taste among these symptoms follows the emergence of evidence suggesting that SARS-CoV-2 frequently impairs the sense of smell. Olfactory disfunction, defined as reduced or distorted ability to smell during sniffing (orthonasal olfaction) or eating (retronasal olfaction), is often reported in mild or even asymptomatic cases [11]. There have also been reports of acute-onset (sudden) anosmia, sometimes in the absence of other symptoms, as a marker of SARS-CoV-2 [12].

Disruption of cells in the olfactory neuroepithelium may result in inflammatory changes that impair olfactory receptor neuron function, cause subsequent olfactory receptor neuron damage, and/or impair subsequent neurogenesis [13]. Such changes may cause temporary or longer-lasting olfactory disease.

Inflammatory signaling molecules are released by infected cells and alveolar macrophages in addition to recruited T lymphocytes, monocytes, and neutrophils. Subsequently the integrity of the alveolar-capillary membrane is compromised by the inflammatory response triggered by SARS-CoV-2 [14]. In the late stage, pulmonary edema can fill the alveolar spaces with hyaline membrane formation, compatible with early-phase acute respiratory distress syndrome [14], bradykinin may contribute to this pulmonary edema [15].

Another contribution for systemic reaction of SARS-CoV-2 infection is the nasal gene expression of ACE-2. Indeed, the lower rates of SARS-CoV-2 infection were found in children. From nasal epithelial samples collected as part of a study involving patients with asthma from 2015 to 2018, a comprehending a cohort of 305 patients aged 4 to 60 years, evidenced that the lower expression of ACE-2 in the nasal epithelium were found in younger children and ACE-2 expression was higher with each subsequent age group after adjusting for sex and asthma [16]. Yet, a recent study bring some data that children may be a potential source of contagion in the SARS-CoV-2 in spite of milder disease or lack of symptoms, and immune dysregulation is implicated in severe post-infectious multisystem inflammatory syndrome in children [17].

3. Implications of angiotensin-converting enzymes and renin-angiotensin system in SARS-CoV-2

Overexpression of human ACE-2 enhanced disease severity of SAR-CoV-2 infection, being the lung injury aggravated by the presence of SARS-CoV spike. Interestingly, in mice model, the lung injury was attenuated by blocking the renin-angiotensin pathway and depended on ACE-2 expression [18].

In contrast to other coronaviruses, SARS-CoV-2 became highly lethal because the virus deregulates a lung protective pathway. About 83% of cells that express ACE-2 were alveolar epithelial type II cells (AECII), suggesting that those cells can serve as a reservoir for viral invasion [19]. In addition, gene ontology enrichment analysis showed that the expression ACE-2 by AECII have high levels of multiple viral process-related genes, including regulatory genes for viral processes, viral life cycle, viral assembly, and viral genome replication, suggesting that the ACE2-expressing AECII facilitate viral replication in the lung [20].

Expression of the ACE-2 receptor is also found in many extrapulmonary tissues including heart, kidney, and intestine [21]. In human lung, the ACE-2 is expressed in endothelial and smooth muscle cells of large and small blood vessels, and in alveolar and bronchial epithelial cells.

Contrarily to ACE-1, the ACE-2 is barely present in the circulation, but widely expressed in mentioned organs. Although ACE-2 is more related to the physiopathology of SARS-CoV, ACE-1 converts angiotensin I into angiotensin Ang II, then ACE-2 break down angiotensin II into molecules that counteract angiotensin II, but if the virus occupies the ACE-2 'receptor' on the surface of cells, then its role is blunted [22]. Angiotensin I, can cause vasoconstriction, inflammation, and fibrosis by signaling through angiotensin II type 1 receptors. ACE-2 cleave angiotensin II to angiotensin 1–7, which can suppress inflammation and fibrosis and generate vasodilation by binding to the *mas* receptor (**Figure 1a**) [23–26].

Moreover, ACE-2 is a negative regulator of the renin-angiotensin system (RAS), and functions as the key SARS coronavirus receptor and stabilizer of neutral amino acid transporters [27]. As previously mentioned, the ACE-2 catalyzes the conversion of angiotensin II to angiotensin 1–7, thereby counterbalancing ACE activity, and converts angiotensin I to generate angiotensin 1–9 [3]. The RAS is an acute phase pathway involved in the multisystemic response of cardiovascular and hematopoietic systems, maintenance of blood pressure homeostasis, as well as fluid and salt balance in mammals [28]. Abnormal activation of RAS has been associated with the pathogenesis of cardiovascular and renal diseases such as hypertension, myocardial infarction and heart failure. Therefore, these disorders share underlying pathophysiology related to the RAS and COVID19 that may be clinically insightful [29].

Cardiovascular disease and pharmacologic RAS inhibition both increase ACE-2 levels, which may increase the virulence of SARS-CoV-2 within the lung and

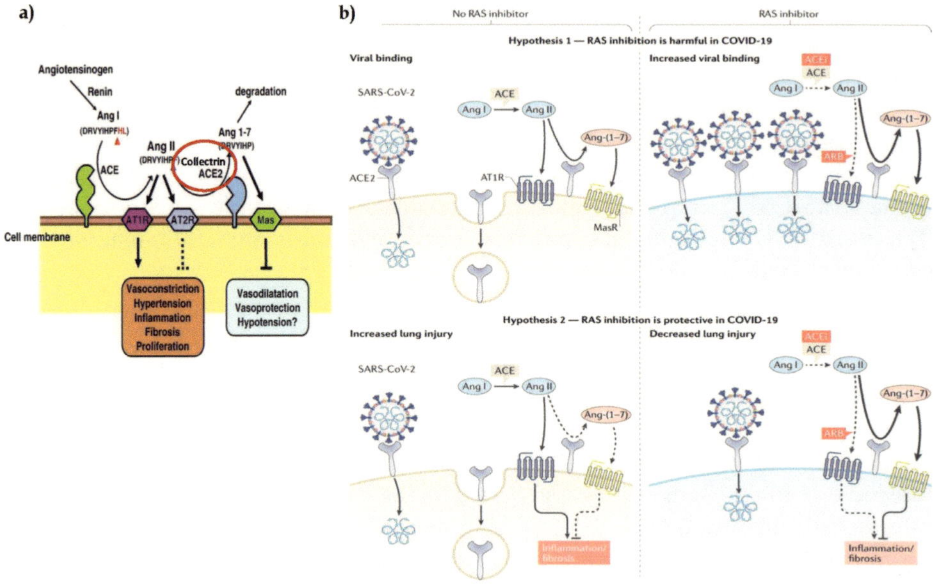

Figure 1.
Integrative schematic diagram of the role of ACE (ACE-1), ACE-2 and collectrin in the renin–angiotensin system (a) (adapted from [30]) and the impact of RAS inhibition in SARS-CoV-2 infection (b) [31].

heart, since the receptor of the two viruses is the same enzyme protein of the cell membrane [32]. Conversely, mechanistic evidence from related coronaviruses suggests that SARS-CoV-2 infection may downregulate ACE-2, leading to toxic over accumulation of angiotensin II that induces acute respiratory distress syndrome and fulminant myocarditis [33]. Therefore, RAS inhibition could mitigate this effect [34]. ACE-2 genetic variants may determine the circulating angiotensin 1–7 levels only in hypertensive females that probably had dose effects related to the localization in the X Chromosome of ACE-2 gene [35].

The bradykinin-kallikrein system can further contribute to local vascular leakage leading to angioedema, due to a local vascular problem because of activation of bradykinin 1 receptor (B1R) and B2R on endothelial cells in the lungs. The RAS is needed to inactivate des-Arg9 bradykinin, which is a potent ligand of the B1R [15]. In the late stage, pulmonary edema can fill the alveolar spaces with hyaline membrane formation, compatible with early-phase acute respiratory distress syndrome.

Other aspect to be pointed out is collectrin (**Figure 1a**), an homolog of ACE-2, that have been identified as essential molecules required for expression of neutral amino acid transporters on the cell surface of epithelial cells. Collectrin (Tmem27) is a transmembrane glycoprotein that is highly expressed in the kidney and vascular endothelium [36]. Furthermore, concordant with metabolic and endocrine changes associated with SARS-CoV-2 infection, collectrin might also have a role in insulin secretion in pancreatic β-cells and/or growth of islet cells [37].

Detailing the mechanism of ACE-1 and its possible role in SARS-CoV-2, ACE-1 has pleiotropic actions involving the cardiovascular and hematopoietic systems [23–25]. The two catalytic domains of ACE-1 has different affinities for its promiscuous substrates respectively in the N domain for goralitide or N-acetyl-seryl-aspartyl-lysyl-proline (NacSDKP), an inhibitor of hematopoiesis and fibrogenesis and that have influence on blood pressure predominantly the C-domain for Angiotensin I or for both domains as is the case of Bradykinin [25, 27].

Unpublished results from our group reflected an inverse correlations of ACE activity with antioxidant erythrocyte and plasma activity enzymes, and direct correlation with lower relative concentrations of glutathione associated to proinflammatory conditions like obesity and several autoimmune diseases (**Figure 2**).

In terms of detection of SARS-CoV-2, the RT-PCR is a cheaper, easier and short turn-around time method for detection of RNA component of SARS-CoV-2, in upper respiratory samples, comparing with sequencing technology. Considering the

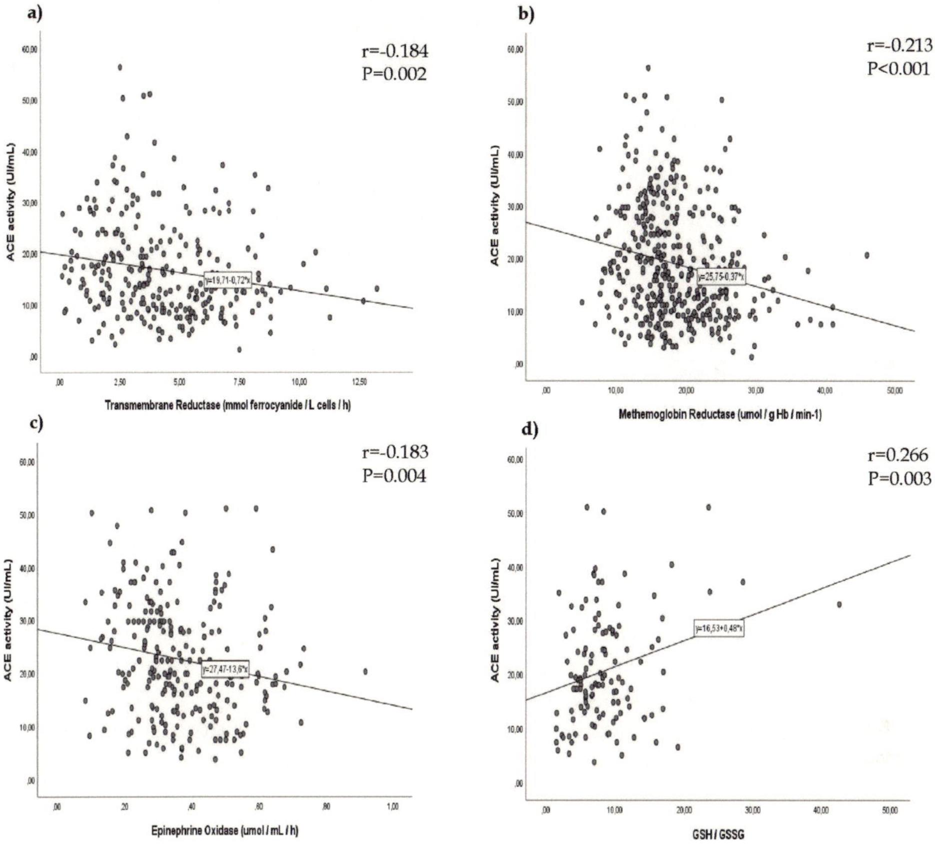

Figure 2.
Correlation between ACE and transmembrane redox system (a), erythrocyte methaemoglobin reductase (b), plasma epinephrine oxidase (c) and with plasma ratio of oxidized glutathione to reduced glutathione (d).

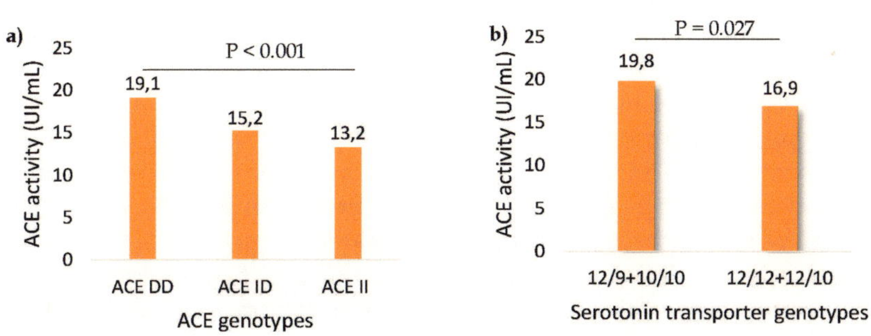

Figure 3.
Distribution of ACE activity according to ACE (a) and SERT (serotonin transporter) (b) genotypes.

genetic variability, the ACE-1 Insertion/Deletion (I/D) functional polymorphism influence its activity in plasma as it was reported by us and other authors (**Figure 3a**) [23, 38]. However, the ACE I/D polymorphism is not associated with increased susceptibility or poor outcome after SARS-CoV-1 infection [39]. Paradoxically, in studies on longevity from our and other groups, individuals with DD genotype, with higher activities of ACE, are more represented in centenarians [40, 41].

The response to this pathway when exaggerated, as is the case of the SARS-CoVs infections, causes intense inflammatory and fibrogenic processes. On the contrary, the system initiated by ACE-2 also has pleiotropic antagonistic actions of the classic system and it has an anti-inflammatory and anti-fibrogenic system [42]. Furthermore, both systems have functional polymorphic genetic variations [23, 38, 39, 43–45].

Genetic polymorphisms in the RAS are putative markers prone to affect the clinical course of SARS-CoV-2 infection. Cao et al. in 2020 suggested that ACE-2 and SARS-CoV-2 associated frequencies among populations can be justified by allele sequences distributions. The greatest are in East Asians populations with higher expressions in tissues that suggest different susceptibilities or response to SARS-CoV-2 in different ecosystems [44].

4. Correlations of immune response, acute phase proteins and tryptophan precursors in SARS-CoV-2 infection

As previously mentioned, the major clinical complication in patients with SARS-CoV-2 is respiratory failure due to local hyperinflammation and acute respiratory distress syndrome. The pathophysiology of these complications has strong similarities to other severe viral lung infections, such as influenza, and other infections caused by coronaviruses (SARS and Middle East respiratory syndrome). An important mechanism mediating lung pathology in these infections is a cytokine storm leading to the so-called "macrophage activation syndrome" with crucial role for monocytes and macrophages [46, 47].

Accordingly with the major clinic complications of this infection, this extreme inflammation compromises the respiratory performance, which often requires ventilator support or, even, extracorporeal membrane oxygenation [48]. However, in approximately 80% of cases, the latter did not prevent mortality, owing to insufficient lung perfusion, which could be explained by developing thromboembolic complications. In this context, clinical trials are underway to determine whether anticoagulants (e.g., heparin) or profibrinolytic drugs (e.g., tissue plasminogen activator) ameliorate severe infection with thromboembolic complications [30, 49].

From the inflammatory perspective, these infection leads to changes in circulating concentrations of proinflammatory cytokines, such as interleukin (IL)-6, tumor necrosis factor (TNF), monocyte chemoattractant protein-1 (MCP-1), macrophage inflammatory protein 1-alpha (MIP1A), and interferon gamma-induced protein 10 (IP10), comparing patients in intensive care unit (ICU) and to those who do not need treatment in the ICU, although the concentrations of some of these cytokines are only moderately increased [50]. This strong increase in systemic inflammation is associated with endothelial dysfunction, increased coagulation activity reflected by elevated d-dimers [50] and hyperactive CCR6 + Th17+ T cells locally in the lung [9]. The increase in systemic concentrations of proinflammatory cytokines was minimal, even during days 7–9, when the patient was symptomatic. This suggests that a mild course of infection is associated with few systemic inflammatory effects. Still, the hyper-inflammation occurs in SARS-CoV-2 and is associated with worse outcomes [48].

Gender differences have been widely discussed in different pathologies, indeed these differences may reflect sex chromosome genes and sex hormones, including estrogens, progesterone, and androgens, with implications to the differential regulation of immune responses between the genders [51]. In studies of hypertension, there is a clear difference between genders taking on account the distribution of ACE-2 genetic polymorphisms associated levels of angiotensin 1-7 [52].

Concerning SARS-CoV-2 infection, a male bias in mortality has emerged in the COVID-19 pandemic, which is consistent with the pathogenesis of other viral infections. Biological gender differences may manifest themselves in susceptibility to infection, early pathogenesis, innate viral control, adaptive immune responses or the balance of inflammation and tissue repair in the resolution of infection [53]. The differences in immune response according with gender, suggest less robust T cell-mediated immunity in male patients with worsening outcome and higher innate cytokine activity, compared to female patients [54].

Evidence reflected the gender as an important driver of risk of mortality and response to the SARS-CoV-2 pandemic. The sex differences in SARS-CoV-2 mortality, severity and recovery, may underly implications of cardiovascular disease (CVD) risk factors, reflecting a plausible biological reasons for this sex difference in SARS-CoV-2 infection [55]. This disproportionate death ratio in men may partly be explained by their relatively higher contribution of pre-existing diseases (i.e., CVD, hypertension, diabetes, and chronic lung disease), higher risk behaviors (i.e., smoking and alcohol use), and occupational exposure [55]. There may be other behavioral and social differences that favor women, with prior studies suggesting women are more likely than men to follow hand hygiene practices and seek preventive care [55].

The host metabolism supports viral pathogenesis by fueling viral proliferation, by providing free amino acids and fatty acids as building blocks. Alterations in tryptophan metabolism and kynurenine pathway regulates inflammation and immunity [56]. The indolamine-2, 3-dioxygenase (IDO) is an intracellular, non-secreted enzyme, which catabolizes kynurenine from tryptophan with interesting role in viral and bacterial infections [57]. Since many microbial organisms rely on the essential amino acid tryptophan, its degradation by IDO-expressing cells of the innate immune system was favored as the major IDO-mediated mechanism against infections [58]. In infectious disease states, IDO has been shown to exert pleiotropic effects, even with opposing outcomes. IDO prevents viral spread and from host perspective also acts to suppress immune reactions thereby promoting infectious diseases [56, 59].

Tryptophan metabolism was the top pathway affected by SARS-CoV-2. As such, focused analysis of this pathway highlighted significant decreases (inversely proportional to IL-6 concentration) in tryptophan, serotonin, and indolepyruvate levels. In contrast, increases in kynurenine, kynurenic acid, picolinic acid, and nicotinic acid suggested hyperactivation of the kynurenine pathway [58]. Furthermore, the levels of IL-6 in serum were significantly different from SARS-Cov-2 patients and controls and they were correlated with changes in tryptophan metabolism [58]. From this study, targeted metabolomics analyses were performed on sera using ultra-high-pressure liquid chromatography-mass spectrometry (UHPLC–MS), highlighting significant associations of COVID-19 and IL-6 levels with amino acid metabolism, purines, acylcarnitines, and fatty acids [58]. Dysregulation of nitrogen metabolism was also seen in infected patients, with altered levels of most amino acids, along with increased markers of oxidant stress (e.g., methionine sulfoxide, cystine), proteolysis, and renal dysfunction (e.g., creatine, creatinine, polyamines). Increased circulating levels of glucose and free fatty acids were also observed, consistent with altered carbon homeostasis. Interestingly, metabolite levels in these pathways correlated with clinical laboratory markers of inflammation (i.e., IL-6 and C-reactive protein) and renal function (i.e., blood urea nitrogen). This initial observational study identified amino acid and fatty acid metabolism as correlates of SARS-CoV-2 [58].

In our group, we also demonstrated that a functional variable number of tandem repeats (VNTR) genetic polymorphism of serotonin transporter, whose expression is activated by IL-1, has some relation with the ACE serum levels that can be associated with unbalanced ACE-ACE-2 system (**Figure 3b**) [38].

Polymorphisms in genes coding for IL-10, TNF-alpha and IL-6 influence circulating levels, and behave as promoters of severe systemic inflammatory response that can probably has an interindividual and gender dependent impact [53].

At the other end of the iceberg, the immunocompromised patients could be protected against SARS-CoV-2, since unlike other common viruses, coronaviruses have not shown to cause more severe disease in immunosuppressed patients, at least statistically significant [60]. Our own immune response appears to be the main driver of lung tissue damage during infection. Starting around the 2nd week of symptoms, patients experience a "storm of cytokines" – autoimmune reaction, where your body over-reacts and in attacking coronavirus, your lungs get caught in the body immunologic response [47, 61]. In the first week of the illness it's the virus itself that's triggering most of your symptoms, but then in severe cases, it's our own inflammatory responses that takes over in causing the most of the damage. So this "storm of cytokines" is killing our immune cells, therefore, could patients with immunosuppressive profile be protected from this reciprocal attack?

The children account for less than 2% of identified cases of SARS-CoV-2 [62]. Interestingly, young children, including infants who are more susceptible to other infections, have milder symptoms and less severe SARS-CoV-2. Nevertheless, children seem to have similar rates of becoming infected compared with middle-aged adults following close contact with a person infected with SARS-CoV-2 [33].

Long-term boosting of innate immune responses, also termed "trained immunity," by certain live vaccines (Bacillus Calmette–Guérin - BCG, oral polio vaccine, measles) induces heterologous protection against infections through epigenetic, transcriptional, and functional reprogramming of innate immune cells [63].

5. Endocrine and metabolic contributions in SARS-CoV-2 infection: ACE-2 downregulation in SARS-CoV-2 as phenocopy of Hartnup disease

Epidemiological data showed that the elderly and those with co-morbidities (diabetes, obesity, and cardiovascular, respiratory, renal, and lung diseases) are most susceptible to COVID-19 and more likely to suffer from the most severe disease complications [64]. Viral infections mobilize free fatty acids to support capsid-associated membrane formation, which was described for other coronaviruses and is explained, in part, by activating phospholipase A2, a target amenable to pharmacological intervention [65].

Hartnup disease is a condition caused by the body's inability to absorb certain protein building blocks (amino acids) from the diet. As a result, affected individuals are not able to use these amino acids to produce other substances, such as vitamins and proteins. Most people with Hartnup disease are able to get the vitamins and other substances they need with a well-balanced diet [27, 66].

Individuals with Hartnup disease have high levels of various amino acids in their urine (aminoaciduria). For most affected individuals, this is the only sign of the condition. However, in other cases, individuals have episodes exhibiting other signs, which can include skin rashes, difficulty of coordination of movements (cerebellar ataxia), and psychiatric symptoms, such as depression or psychosis. These episodes are typically temporary and are often triggered by intercurrent infection, stress, nutrient-poor diet, or fever. These features tend to go away once the trigger is changed, although the aminoaciduria remains. In affected individuals, signs and symptoms most commonly occur in childhood [67, 68].

As previously mentioned, the two antagonistic systems ACE, ANG II, AT1R and ACE2, ANII 1–7 are in the "hurricane eye" of SARS-CovV-2 and the non-enzymatic role of ACE-2 give rise to Hartnup disease phenocopy. ACE-2 is also a stabilizing protein (very similar to collectrin in kidney) of the neutral amino acid transporter mutated in the Hartnup disease [27].

In mice with ACE-2 deletion in the small intestine, there was also a decrease in tryptophan absorption secondary to the lower expression of the neutral amino acid transporter accompanied by a phenotype very similar to that of Hartnup's disease phenotypes [69]. This situation can be caused by SARS-COVs and probably explains the gastro intestinal symptoms sometimes associated with those viral infections. In this case, it may be the result of the accumulation of nephrotoxic and pro-inflammatory pulmonary products (indole derivatives) or lack of anti-inflammatory kynurenines (IDO derivatives), as a consequence of dysbiosis at large intestine resulting from the lack of absorption of several neutral and aromatic amino acids namely tryptophan [70, 71].

6. New highlights of possible microbioma association to SARS-CoV-2 infection

Concordantly to exposed in this chapter, the SARS-CoV-2 is more than a severe respiratory infection and actually integrate a multisystemic coordination. Metabolic syndrome and microbiome had been associated in intervention from ACE-2. This relation has an explanation that is now much more clarified and that goes through the IDO derivatives (Kynurenines) associated with aryl hydrocarbon receptor (AhR) and anti-inflammatory response Th22 [56].

The rationale of the non-enzymatic role of ACE-2 to serotonin and IDO derivatives to kynurenines has an explanation based in the activation of AhR functions by these tryptophan metabolites as they activates anti-inflammatory cytokines that may counteract the SARS-CoV-2 gastrointestinal and pulmonary symptoms characterized by a "cytokine storm" [72]. This can have their origin in the dysbiosis related to the tryptophan catabolism in indol derivatives by unbalanced *Lacobacillus spp* (decreased) specially in high salt microenvironment characteristic of western pattern diets [71, 73, 74].

Importantly, ACE-2 is highly expressed on the luminal surface of intestinal epithelial cells, functioning as a co-receptor for nutrient uptake, in particular for amino acid resorption from food [75]. Therefore the intestine might also be a major entry site for SARS-CoV-2 and the infection might have been initiated by eating food from the Wuhan market, the putative site of the outbreak. Whether SARS-CoV-2 can indeed infect the human gut epithelium has important implications for fecal–oral transmission and containment of viral spread. Moreover, the ACE-2 tissue distribution in other organs could explain the multi-organ dysfunction observed in patients [66, 71, 76, 77]. Any perturbation in host-microbiota crosstalk can be an initiating or re-enforcing factor in SARS-CoV-2 pathogenesis.

Some bacteria produce bioactive neurotransmitters that have previously been proposed to modulate nervous system activity and behaviors of their host. A large array of metabolites drives the crosstalk between the host and its microbiome. The three currently most studied categories of metabolites involved in host-microbiota interactions are short-chain fatty acids produced by bacteria from the fermentation of fibbers, bile acids produced in the liver and transformed by the gut microbiota before re-affecting the host, and tryptophan metabolites, which are the topic of this review [72].

Tryptophan is an essential aromatic amino acid composed of a b carbon connected to the 3 position of an indole group and it is a biosynthetic precursor of a large number of microbial and host metabolites [78]. It's metabolism follows three major pathways in the gastrointestinal tract: the direct transformation of Tryptophan into several molecules, including ligands of the (AhR) by the gut microbiota [78]; the kynurenine pathway in both immune and epithelial cells via IDO-1 [79]; and the serotonin (5-hydroxytryptamine [5-HT]) production pathway in enterochromaffin cells via Tryptophan hydroxylase 1 (TpH1) [72]. The AhR is implicated in lung inflammation [80].

The gut microbiota influences the health of the host, especially with regard to gut immune homeostasis and the intestinal immune response. In addition to serving as a nutrient enhancer, L-tryptophan plays crucial roles in the balance between intestinal immune tolerance and gut microbiota maintenance.

7. Final marks

These lessons derived of SARS-CoVs infections outbreaks (2003 and 2019) can explain the role of the two antagonistic RASs pathways on the hypoxic pulmonary vasoconstriction an homeostatic mechanism in response to alveolar hypoxia secondary to acute lung injury in SARS, optimizing ventilation, perfusion and systemic oxygen delivery. Moreover, the new knowledge about the role of RAS proteins, namely, ACE-2 in gut with pleiotropic actions on the metabolism of tryptophan in the crosstalk microbiota–intestine, intestine-kidney and probably intestine-lung can help in designing new, based on probiotics and prebiotics or repurposing ancient therapies for disorders involving those organ crosstalk resultant physio pathologies.

Acknowledgements

The authors would like to acknowledge the Instituto de Investigação Científica Bento da Rocha Cabral and Sociedade Portuguesa de Papillomavírus for support.

Conflict of interest

The authors declare that they have no competing interests.

Author details

Andreia Matos[1,2,3,4*], Alda Pereira da Silva[5,6], Joana Ferreira[1,2], Ana Carolina Santos[1,2], Maria Clara Bicho[7,8] and Manuel Bicho[1,2]

1 Genetics Laboratory and ISAMB - Environmental Health Institute - Ecogenetics and Human Health Unit, Faculty of Medicine of University of Lisbon, Av. Prof. Egas Moniz, Piso 1C, 1649-028 Lisboa, Portugal

2 Instituto de Investigação Científica Bento da Rocha Cabral, Lisbon, Calçada Bento da Rocha Cabral, n°14, 1250-047, Portugal

3 Tumor and Microenvironment Interactions Group, i3S - Instituto de Investigação e Inovação em Saúde/INEB-Institute of Biomedical Engineering, University of Porto, Rua Alfredo Allen, 208, 4200-135 Porto, Portugal

4 Abel Salazar Institute for the Biomedical Sciences (ICBAS), University of Porto, Rua de Jorge Viterbo Ferreira No. 228, 4050-313 Porto, Portugal

5 ISAMB - Environmental Health Institute - Ecogenetics and Human Health Unit, University Clinic of General and Family Medicine, Institute for Preventive Medicine and Public Health, Faculty of Medicine, University of Lisbon, Av. Professor Egas Moniz, Piso 1C 1649-028, Lisboa, Portugal

6 CBIOS - Biosciences Research Center, School of Health Sciences and Technologies, Universidade Lusófona, Campo Grande 376,1649-024, Lisboa, Portugal

7 Institute of Preventive Medicine and Public Health and Genetics Laboratory and ISAMB - Environmental Health Institute of University of Lisbon Faculty of Medicine; Av. Prof. Egas Moniz, Piso 1C, 1649-028, Lisboa, Portugal

8 Dermatology Research Unit, Instituto de Medicina Molecular, Av. Prof. Egas Moniz, 1649-028 Lisboa, Portugal

*Address all correspondence to: andreiamatos@medicina.ulisboa.pt

References

[1] Andersen KG, Rambaut A, Lipkin WI, Holmes EC, Garry RF. The proximal origin of SARS-CoV-2. Nat Med. 2020;26(4): 450-2.

[2] Jaimes JA, André NM, Chappie JS, Millet JK, Whittaker GR. Phylogenetic Analysis and Structural Modeling of SARS-CoV-2 Spike Protein Reveals an Evolutionary Distinct and Proteolytically Sensitive Activation Loop. J Mol Biol. 2020;432(10):3309-25.

[3] Kuba K, Imai Y, Rao S, Gao H, Guo F, Guan B, et al. A crucial role of angiotensin converting enzyme 2 (ACE2) in SARS coronavirus-induced lung injury. Nat Med. 2005;11(8):875-9.

[4] Hoffmann M, Kleine-Weber H, Schroeder S, Krüger N, Herrler T, Erichsen S, et al. SARS-CoV-2 Cell Entry Depends on ACE2 and TMPRSS2 and Is Blocked by a Clinically Proven Protease Inhibitor. Cell. 2020;181(2):271-80.e8.

[5] Shulla A, Heald-Sargent T, Subramanya G, Zhao J, Perlman S, Gallagher T. A transmembrane serine protease is linked to the severe acute respiratory syndrome coronavirus receptor and activates virus entry. J Virol. 2011;85(2):873-82.

[6] Catanzaro M, Fagiani F, Racchi M, Corsini E, Govoni S, Lanni C. Immune response in COVID-19: addressing a pharmacological challenge by targeting pathways triggered by SARS-CoV-2. Signal Transduct Target Ther. 2020;5(1):84.

[7] Netea MG, Domínguez-Andrés J, Barreiro LB, Chavakis T, Divangahi M, Fuchs E, et al. Defining trained immunity and its role in health and disease. Nat Rev Immunol. 2020;20(6):375-88.

[8] Schäfer A, Baric RS. Epigenetic Landscape during Coronavirus Infection. Pathogens. 2017;6(1):8.

[9] Xu X, Yu C, Qu J, Zhang L, Jiang S, Huang D, et al. Imaging and clinical features of patients with 2019 novel coronavirus SARS-CoV-2. Eur J Nucl Med Mol Imaging. 2020;47(5):1275-80.

[10] Lin L, Jiang X, Zhang Z, Huang S, Zhang Z, Fang Z, et al. Gastrointestinal symptoms of 95 cases with SARS-CoV-2 infection. Gut. 2020;69(6):997-1001.

[11] Whitcroft KL, Hummel T. Olfactory Dysfunction in COVID-19: Diagnosis and Management. Jama. 2020.

[12] Whitcroft KL, Hummel T. Olfactory Dysfunction in COVID-19: Diagnosis and Management. JAMA. 2020;323(24):2512-4.

[13] Li Y-C, Bai W-Z, Hashikawa T. The neuroinvasive potential of SARS-CoV2 may play a role in the respiratory failure of COVID-19 patients. J Med Virol. 2020;92(6):552-5.

[14] Wiersinga WJ, Rhodes A, Cheng AC, Peacock SJ, Prescott HC. Pathophysiology, Transmission, Diagnosis, and Treatment of Coronavirus Disease 2019 (COVID-19): A Review. Jama. 2020.

[15] van de Veerdonk FL, Netea MG, van Deuren M, van der Meer JW, de Mast Q, Brüggemann RJ, et al. Kallikrein-kinin blockade in patients with COVID-19 to prevent acute respiratory distress syndrome. Elife. 2020;9.

[16] Bunyavanich S, Do A, Vicencio A. Nasal Gene Expression of Angiotensin-Converting Enzyme 2 in Children and Adults. Jama. 2020;323(23):2427-9.

[17] Yonker LM, Neilan AM, Bartsch Y, Patel AB, Regan J, Arya P, et al. Pediatric SARS-CoV-2: Clinical Presentation, Infectivity, and Immune Responses. The Journal of Pediatrics.

[18] Imai Y, Kuba K, Rao S, Huan Y, Guo F, Guan B, et al. Angiotensin-converting enzyme 2 protects from severe acute lung failure. Nature. 2005;436(7047):112-6.

[19] Zhao Y, Zhao Z, Wang Y, Zhou Y, Ma Y, Zuo W. Single-cell RNA expression profiling of ACE2, the receptor of SARS-CoV-2. bioRxiv. 2020:2020.01.26.919985.

[20] Zhang H, Penninger JM, Li Y, Zhong N, Slutsky AS. Angiotensin-converting enzyme 2 (ACE2) as a SARS-CoV-2 receptor: molecular mechanisms and potential therapeutic target. Intensive Care Med. 2020;46(4):586-90.

[21] Li MY, Li L, Zhang Y, Wang XS. Expression of the SARS-CoV-2 cell receptor gene ACE2 in a wide variety of human tissues. Infect Dis Poverty. 2020;9(1):45.

[22] Sriram K, Insel PA. A hypothesis for pathobiology and treatment of COVID-19: The centrality of ACE1/ACE2 imbalance. Br J Pharmacol. 2020.

[23] Almeida SS, Barros CC, Moraes MR, Russo FJ, Haro AS, Rosa TS, et al. Plasma Kallikrein and Angiotensin I-converting enzyme N- and C-terminal domain activities are modulated by the insertion/deletion polymorphism. Neuropeptides. 2010;44(2):139-43.

[24] Bernstein KE, Shen XZ, Gonzalez-Villalobos RA, Billet S, Okwan-Duodu D, Ong FS, et al. Different in vivo functions of the two catalytic domains of angiotensin-converting enzyme (ACE). Curr Opin Pharmacol. 2011;11(2):105-11.

[25] Kanasaki K. N-acetyl-seryl-aspartyl-lysyl-proline is a valuable endogenous antifibrotic peptide for kidney fibrosis in diabetes: An update and translational aspects. Journal of Diabetes Investigation. 2020;11(3):516-26.

[26] Luft FC. High-mobility group box 1 protein, angiotensins, ACE2, and target organ damage. J Mol Med (Berl). 2016;94(1):1-3.

[27] Kuba K, Imai Y, Ohto-Nakanishi T, Penninger JM. Trilogy of ACE2: a peptidase in the renin-angiotensin system, a SARS receptor, and a partner for amino acid transporters. Pharmacol Ther. 2010;128(1):119-28.

[28] Sparks MA, Crowley SD, Gurley SB, Mirotsou M, Coffman TM. Classical Renin-Angiotensin system in kidney physiology. Compr Physiol. 2014;4(3):1201-28.

[29] Guzik TJ, Mohiddin SA, Dimarco A, Patel V, Savvatis K, Marelli-Berg FM, et al. COVID-19 and the cardiovascular system: implications for risk assessment, diagnosis, and treatment options. Cardiovasc Res. 2020;116(10):1666-87.

[30] Zhang Q, Chen CZ, Swaroop M, Xu M, Wang L, Lee J, et al. Targeting heparan sulfate proteoglycan-assisted endocytosis as a COVID-19 therapeutic option. bioRxiv. 2020:2020.07.14.202549.

[31] South AM, Tomlinson L, Edmonston D, Hiremath S, Sparks MA. Controversies of renin-angiotensin system inhibition during the COVID-19 pandemic. Nat Rev Nephrol. 2020;16(6):305-7.

[32] Gheblawi M, Wang K, Viveiros A, Nguyen Q, Zhong JC, Turner AJ, et al. Angiotensin-Converting Enzyme 2: SARS-CoV-2 Receptor and Regulator of the Renin-Angiotensin System: Celebrating the 20th Anniversary of the Discovery of ACE2. Circ Res. 2020;126(10):1456-74.

[33] Bi Q, Wu Y, Mei S, Ye C, Zou X, Zhang Z, et al. Epidemiology and transmission of COVID-19 in 391 cases and 1286 of their close contacts in Shenzhen, China: a retrospective cohort study. Lancet Infect Dis. 2020.

[34] Rico-Mesa JS, White A, Anderson AS. Outcomes in Patients with COVID-19 Infection Taking ACEI/ARB. Curr Cardiol Rep. 2020;22(5):31.

[35] Chen YY, Zhang P, Zhou XM, Liu D, Zhong JC, Zhang CJ, et al. Relationship between genetic variants of ACE2 gene and circulating levels of ACE2 and its metabolites. J Clin Pharm Ther. 2018;43(2):189-95.

[36] Kuba K, Imai Y, Ohto-Nakanishi T, Penninger JM. Trilogy of ACE2: a peptidase in the renin-angiotensin system, a SARS receptor, and a partner for amino acid transporters. Pharmacology & therapeutics. 2010;128(1):119-28.

[37] Graus-Nunes F, Souza-Mello V. The renin-angiotensin system as a target to solve the riddle of endocrine pancreas homeostasis. Biomedicine & Pharmacotherapy. 2019;109:639-45.

[38] Marinho C, Ferreira J, Levy P, Coelho C, Gomes A, Lares M, et al. Contribution of genetic polymorphisms involved in the control of blood pressure and cell growth/oxidative stress for the modulation of angiotensin converting enzyme activity. Portuguese Journal of Cardiology. 2011;30(2):189-95.

[39] Chan KC, Tang NL, Hui DS, Chung GT, Wu AK, Chim SS, et al. Absence of association between angiotensin converting enzyme polymorphism and development of adult respiratory distress syndrome in patients with severe acute respiratory syndrome: a case control study. BMC Infect Dis. 2005;5:26.

[40] Garatachea N, Marín PJ, Lucia A. The ACE DD genotype and D-allele are associated with exceptional longevity: a meta-analysis. Ageing Res Rev. 2013;12(4):1079-87.

[41] Pereira da Silva A, Matos A, Aguiar L, Ramos-Marques N, Ribeiro R, Gil Â, et al. Hypertension and longevity: role of genetic polymorphisms in renin-angiotensin-aldosterone system and endothelial nitric oxide synthase. Mol Cell Biochem. 2019;455(1-2):61-71.

[42] Luft FC. The renin-angiotensin system and prolylcarboxypeptidase. J Mol Med (Berl). 2017;95(5):461-3.

[43] Adamzik M, Frey U, Sixt S, Knemeyer L, Beiderlinden M, Peters J, et al. <em>ACE</em> I/D but not <em>AGT</em> (−6)A/G polymorphism is a risk factor for mortality in ARDS. European Respiratory Journal. 2007;29(3):482.

[44] Cao Y, Li L, Feng Z, Wan S, Huang P, Sun X, et al. Comparative genetic analysis of the novel coronavirus (2019-nCoV/SARS-CoV-2) receptor ACE2 in different populations. Cell Discov. 2020;6:11.

[45] Zhang Q, Cong M, Wang N, Li X, Zhang H, Zhang K, et al. Association of angiotensin-converting enzyme 2 gene polymorphism and enzymatic activity with essential hypertension in different gender: A case-control study. Medicine (Baltimore). 2018;97(42):e12917.

[46] Astuti I, Ysrafil. Severe Acute Respiratory Syndrome Coronavirus 2 (SARS-CoV-2): An overview of viral structure and host response. Diabetes & metabolic syndrome. 2020;14(4):407-12.

[47] Mehta P, McAuley DF, Brown M, Sanchez E, Tattersall RS, Manson JJ. COVID-19: consider cytokine storm syndromes and immunosuppression. Lancet. 2020;395(10229):1033-4.

[48] Wang D, Chai XQ, Magnussen CG, Zosky GR, Shu SH, Wei X, et al. Renin-angiotensin-system, a potential pharmacological candidate, in acute respiratory distress syndrome

during mechanical ventilation. Pulm Pharmacol Ther. 2019;58:101833.

[49] Belen-Apak FB, Sarialioglu F. The old but new: Can unfractioned heparin and low molecular weight heparins inhibit proteolytic activation and cellular internalization of SARS-CoV2 by inhibition of host cell proteases? Med Hypotheses. 2020;142:109743.

[50] Zhou F, Yu T, Du R, Fan G, Liu Y, Liu Z, et al. Clinical course and risk factors for mortality of adult inpatients with COVID-19 in Wuhan, China: a retrospective cohort study. Lancet. 2020;395(10229):1054-62.

[51] Scully EP, Haverfield J, Ursin RL, Tannenbaum C, Klein SL. Considering how biological sex impacts immune responses and COVID-19 outcomes. Nat Rev Immunol. 2020;20(7):442-7.

[52] Liu D, Chen Y, Zhang P, Zhong J, Jin L, Zhang C, et al. Association between circulating levels of ACE2-Ang-(1-7)-MAS axis and ACE2 gene polymorphisms in hypertensive patients. Medicine (Baltimore). 2016;95(24):e3876.

[53] Lefèvre N, Corazza F, Valsamis J, Delbaere A, De Maertelaer V, Duchateau J, et al. The Number of X Chromosomes Influences Inflammatory Cytokine Production Following Toll-Like Receptor Stimulation. Frontiers in Immunology. 2019;10(1052).

[54] Park MD. Sex differences in immune responses in COVID-19. Nat Rev Immunol. 2020;20(8):461.

[55] Sharma G, Volgman AS, Michos ED. Sex Differences in Mortality From COVID-19 Pandemic. JACC: Case Reports. 2020;2(9):1407.

[56] Costantini C, Bellet MM, Renga G, Stincardini C, Borghi M, Pariano M, et al. Tryptophan Co-Metabolism at the Host-Pathogen Interface. Frontiers in immunology. 2020;11:67-.

[57] Schmidt SV, Schultze JL. New Insights into IDO Biology in Bacterial and Viral Infections. Front Immunol. 2014;5:384.

[58] Thomas T, Stefanoni D, Reisz JA, Nemkov T, Bertolone L, Francis RO, et al. COVID-19 infection alters kynurenine and fatty acid metabolism, correlating with IL-6 levels and renal status. JCI Insight. 2020;5(14).

[59] Davies NWS, Guillemin G, Brew BJ. Tryptophan, Neurodegeneration and HIV-Associated Neurocognitive Disorder. International journal of tryptophan research : IJTR. 2010;3:121-40.

[60] Gao Y, Chen Y, Liu M, Shi S, Tian J. Impacts of immunosuppression and immunodeficiency on COVID-19: A systematic review and meta-analysis. J Infect. 2020;81(2):e93-e5.

[61] Nicholls J, Peiris M. Good ACE, bad ACE do battle in lung injury, SARS. Nat Med. 2005;11(8):821-2.

[62] Wu Z, McGoogan JM. Characteristics of and Important Lessons From the Coronavirus Disease 2019 (COVID-19) Outbreak in China: Summary of a Report of 72 314 Cases From the Chinese Center for Disease Control and Prevention. Jama. 2020.

[63] Sánchez-Ramón S, Conejero L, Netea MG, Sancho D, Palomares Ó, Subiza JL. Trained Immunity-Based Vaccines: A New Paradigm for the Development of Broad-Spectrum Anti-infectious Formulations. Frontiers in Immunology. 2018;9(2936).

[64] Sanyaolu A, Okorie C, Marinkovic A, Patidar R, Younis K, Desai P, et al. Comorbidity and its

Impact on Patients with COVID-19. SN Compr Clin Med. 2020:1-8.

[65] Ma S, Mao Q, Chen W, Zhao M, Wu K, Song D, et al. Serum Lipidomics Analysis of Classical Swine Fever Virus Infection in Piglets and Emerging Role of Free Fatty Acids in Virus Replication in vitro. Frontiers in Cellular and Infection Microbiology. 2019;9(410).

[66] Perlot T, Penninger JM. ACE2 - from the renin-angiotensin system to gut microbiota and malnutrition. Microbes Infect. 2013;15(13):866-73.

[67] Baron DN, Dent CE, Harris H, Hart EW, Jepson JB. Hereditary pellagra-like skin rash with temporary cerebellar ataxia, constant renal amino-aciduria, and other bizarre biochemical features. Lancet. 1956;271(6940):421-8.

[68] Seow HF, Bröer S, Bröer A, Bailey CG, Potter SJ, Cavanaugh JA, et al. Hartnup disorder is caused by mutations in the gene encoding the neutral amino acid transporter SLC6A19. Nat Genet. 2004;36(9):1003-7.

[69] Malakauskas SM, Quan H, Fields TA, McCall SJ, Yu MJ, Kourany WM, et al. Aminoaciduria and altered renal expression of luminal amino acid transporters in mice lacking novel gene collectrin. Am J Physiol Renal Physiol. 2007;292(2):F533-44.

[70] Lu CC, Ma KL, Ruan XZ, Liu BC. Intestinal dysbiosis activates renal renin-angiotensin system contributing to incipient diabetic nephropathy. Int J Med Sci. 2018;15(8):816-22.

[71] Müller DN, Wilck N, Haase S, Kleinewietfeld M, Linker RA. Sodium in the microenvironment regulates immune responses and tissue homeostasis. Nat Rev Immunol. 2019;19(4):243-54.

[72] Agus A, Planchais J, Sokol H. Gut Microbiota Regulation of Tryptophan Metabolism in Health and Disease. Cell Host Microbe. 2018;23(6):716-24.

[73] Dhar D, Mohanty A. Gut microbiota and Covid-19- possible link and implications. Virus Res. 2020;285:198018.

[74] Kalantar-Zadeh K, Ward SA, Kalantar-Zadeh K, El-Omar EM. Considering the Effects of Microbiome and Diet on SARS-CoV-2 Infection: Nanotechnology Roles. ACS Nano. 2020;14(5):5179-82.

[75] Zhang H, Li H-B, Lyu J-R, Lei X-M, Li W, Wu G, et al. Specific ACE2 expression in small intestinal enterocytes may cause gastrointestinal symptoms and injury after 2019-nCoV infection. International Journal of Infectious Diseases. 2020;96:19-24.

[76] Dhar D, Mohanty A. Gut microbiota and Covid-19- possible link and implications. Virus Research. 2020;285:198018.

[77] Gao QY, Chen YX, Fang JY. 2019 Novel coronavirus infection and gastrointestinal tract. Journal of Digestive Diseases. 2020;21(3):125-6.

[78] Alkhalaf LM, Ryan KS. Biosynthetic manipulation of tryptophan in bacteria: pathways and mechanisms. Chem Biol. 2015;22(3):317-28.

[79] Clarke G, Fitzgerald P, Cryan JF, Cassidy EM, Quigley EM, Dinan TG. Tryptophan degradation in irritable bowel syndrome: evidence of indoleamine 2,3-dioxygenase activation in a male cohort. BMC Gastroenterol. 2009;9:6-.

[80] Beamer CA, Shepherd DM. Role of the aryl hydrocarbon receptor (AhR) in lung inflammation. Semin Immunopathol. 2013;35(6):693-704.

Chapter 5

Ebola, the Negative Stranded RNA Virus

Aqsa Farman, Syed Lal Badshah, Khalid Khan, Nasir Ahmad and Abdul Naeem

Abstract

Ebola virus (EBV) is a deadly virus that has resulted in a number of deaths during its outbreaks in Africa in 2014–2016 and 2018–2019. This virus causes a hemorrhagic fever like other pathogenic viruses of the Filoviridae family with high mortality rate. The exact reservoir of the ebola virus is not known, but different mammal groups are the source from which it is transferred to the human population. The transmission among the human population is through body fluids of patients and also through aersol droplets in the air. The role of different glycoproteins in the budding formation has helped a lot in understanding the physiology of the ebola virus. Most of these viral glycoproteins synthesis and the replication enzymes offer a good inhibitory target for drug design against the ebola virus. Recently, different groups have claimed the development of a successful vaccine for the ebola virus. However, the availability of the vaccines to the poor population of Africa and other parts of the world is still not practical.

Keywords: Ebola virus, hemorrhagic fever, vaccine, glycoproteins, molecular docking

1. Introduction and background

The Ebola virus was first discovered back in 1976 when its breakout occurred simultaneously in South Sudan and in the area of Ebola River in Yambuku city of Democratic Republic of Congo (**Figure 1**) [1]. It reemerges in 1990s, 2000s, peaked in 2014 and 2018–2019 (**Figure 1**). This RNA viral disease caused hemorrhagic fever in humans and non-human primates [2]. The African originated species of Ebola virus fatality rate in humans is nearly 90% [2]. Ebola, Marburg, and Cueva virus are the three infectious viruses with a common ancestor of the family Filoviridae [2]. The word "filo" in Filoviridae is derived from the Latin word film, which means thread like, as these virions appear thread-like under the electron microscope [3]. The Filoviridae family is part of the order Mononegavirales, class Monjiviricetes and phylum Negarnaviricota in the virus taxonomic classification [4, 5]. Ebola, also known as Ebola virus disease (EVD/EBOV) and Ebola hemorrhagic fever (EHF) in the past, is a filamentous, enveloped, non-segmented, single stranded negative sense ribonucleic acid (RNA) virus [4]. The replication of Filoviridae family of viruses takes place in the cytoplasm of the host cell [6, 7]. The genomic RNA is non-contagious alone because it is not able to serve as a template for protein synthesis [6, 7]. To start the transcription of positive-sense messenger RNA (mRNA) the viral protein must connect with the genomic RNA [6, 7].

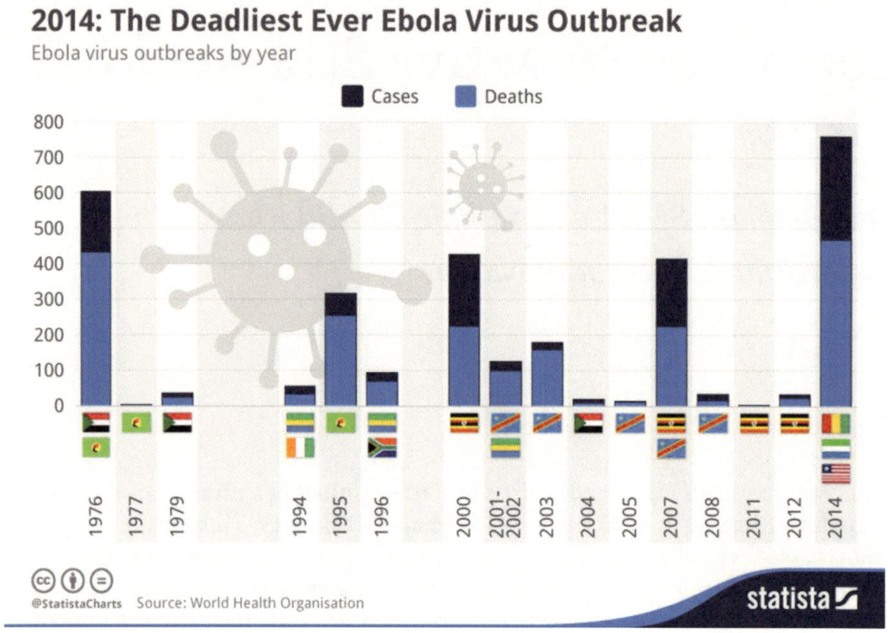

Figure 1.
Reported cases and deaths from EBOV in different countries from 1976 to 2014 according to WHO [8].

The objective of this work is to provide firsthand information regarding the EBOV origin, infection, control and progress in vaccine development.

1.1 Origin

The Ebola virus is taxonomically classified into five species. These species includes Zaire Ebola virus (EBOV), Reston Ebola virus (RESTV), Taï Forest Ebola virus (TAFV) or Coted'Ivoire Ebola virus (CIEBOV) [9], Sudan Ebola virus (SUDV), and Bundibugyo Ebola virus (BDBV) [10]. Four of these species are responsible to cause disease in humans and one in nonhuman primates [10]. The species of Ebola viruses are named from where they originated for the first time [11]. The BDBV was first found in 2007 in Bundibugyo district of Uganda with 116 cases and fatality rate of 25% [12, 13]. RESTV emerged in 1989 in Reston (Virginia), USA, causing diseases in nonhuman primates [14, 15]. SUDV was first discovered in 1976 in a cotton factory in South Sudan (Nzara). Its mortality rate varies from 53 to 68%. The mortality rate is the death rate due to certain disease in a particular population. CIEBOV (Tai forest virus) was first reported in 1994. The virus appeared in chimpanzees in Africa in Tai forest [16]. ZEBOV has the highest lethality rate up to 90%, which includes the first appeared virus and the most current outbreak [17].

1.2 History

The Ebola virus is named after a small river, Ebola in Zaire and this viral infection first appeared in 1976 in both South Sudan and Zaire in the African continent [18–20]. The international teams of WHO reached the affected area but were unsuccessful in saving more lives and collecting vital information due to lack of medical facilities and illiteracy in rural areas [21]. This virus family becomes genetically more advanced and spread into other areas, and appeared in the united states in 1989 [22]. EBOV did not appear in Africa for 15 years but as the natural reservoir was not completely eliminated and due to this reason the virus showed itself again in 1994–1996 [23]. During 1994–1996, the new subtype of Ebola Coted'Ivoire was discovered, while the old subtypes were also present. The people infected were living near the tropical forest area. The ethnologist's discovered that the fifth subtype was found in the chimpanzee group with a high fatality rate. They were transferred to Switzerland for suitable facilities and medical care by a group of scientists [24]. The outbreak in 2014 has the highest fatality rates in the history of EBOV [25].

1.3 Transmission, symptoms, and treatment

EBOV is a rare and deadly disease caused by infection with one of the Ebola virus strains. Humans and non-human primates are the primary hosts of Ebola [26]. The different types of insectivorous bat genus, especially *Mops condylurus* and frugivorous bat species are also the host of EBOV [27–29]. The fruits eating bat is also a common reservoir of ebola virus [30, 31]. The virus is transmitted through contaminated blood and bodily fluids of infected person or animal. Fever, muscle ache, loss of appetite, and fatigue are the common symptoms after onset of EBOV, while in fatal cases, severe vomiting, diarrhea, hemorrhage, septic shock, and multi-organ failure may occur that may lead to death in 30–90% patients [32]. The patients can be treated as the symptoms of EBOV appear in them. Currently, there is no proper Federal Drug Authority (FDA) approved medicines or vaccines. The following are some of the interventions and precautions for survival against EBOV;

a. Intravenous fluids (IV) and body salts should be regularly provided to the EBOV infected patients.

b. The blood pressure and oxygen status of patients should be monitored and maintained to the normal level.

c. The blood lost during the illness should be replaced and medicines should be provided to control the blood loss.

d. The patient should increase the intake of fluids.

e. The other infections which develop during treatment should also be treated simultaneously [33].

1.4 EBOV genomic organizations

The EBOV genome has nucleotide sequences which are extragenic. The sequences combine to produce secondary structures that contain 3′ and 5′ ends. The genome serves to initiate transcription and genome replication [34, 35]. The negative-stranded RNA genome of EBOV contains seven important genes that are arranged in a linear fashion and has a total length of approximately 19 kilobases and it translated into eight proteins [36]. Its sequence is as follows 3′-NP, VP35, VP40, GP, VP30, VP24, L-5′as shown in **Figure 2**. The VP (35) is RNA-dependent RNA polymerase cofactor, VP(40) is a matrix protein, GP(1,2) is spike glycoprotein, VP(30) is a transcriptional activator, VP (24) is the second protein matrix and (L) is RNA polymerase enzyme [6]. The RNA genome of the virus is coated by a complex of ribonuclear protein (RNP). It is also known as nucleocapsid. The nucleocapsid is composed of major nucleoprotein (NP), minor nucleoproteins that include VP(30), VP(35), and polymerase (L) protein as shown in **Figure 3** [37, 38]. The above complex is then enveloped in a layer of matrix. The matrix consists of the major and minor matrix protein VP40, VP24, respectively. Then, the matrix proteins are encapsulated by a double layer of lipid which is then peppered with the viral glycoprotein (GP) that help in binding to the host cell [39, 40]. All five EBOV species discovered so far has the same GP gene organization which is divided into sGP, GP1, GP2, and ssGP [41, 42].

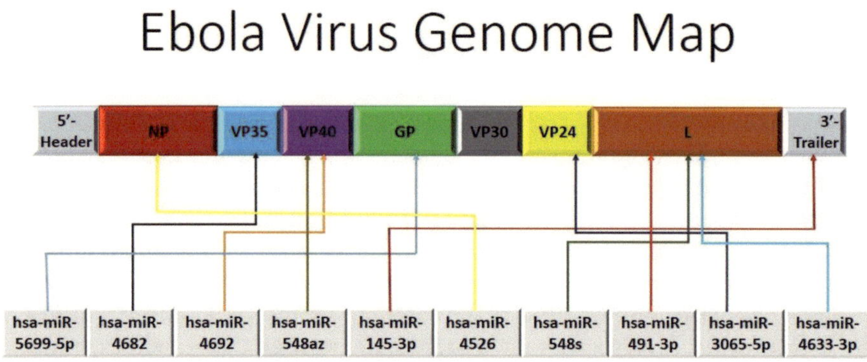

Figure 2.
Important proteins in the RNA genome of EBOV [43].

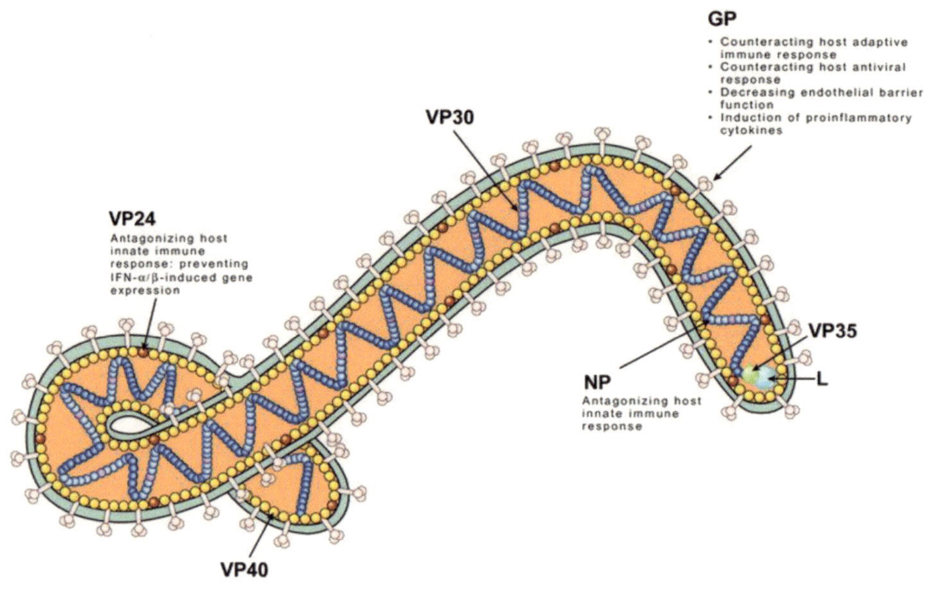

Figure 3.
The virion structure showing the nuclear material and glycoproteins [44].

1.5 Structure of glycoproteins

It has been observed that most of the viral components play an active role in the pathogenesis and infection of this viral disease, but the major role is played by the glycoproteins of the EBOV [45]. Thus, they possess a significant importance in the virulence of EBOV and targeting its synthesis is one of important step for controlling this infection [45]. EBOV includes different glycoproteins each of which plays their roles in different aspects of viral life cycle. Each gene product has distinct biochemical and biological properties [46, 47]. The fourth gene, GP is formed by a process known as glycosylation in which carbohydrate are linked covalently to the polypeptide chain [48].

1.6 Production of glycoproteins

The biosynthesis of glycoproteins occurs by transcription and expression strategies. The precursor of the secreted glycoproteins and full length glycoproteins are the initial products of EBOV glycoprotein. Pre-sGP is transcripted and translated to soluble products sGP and Δ-peptide while glycoprotein GP1, 2 is cleaved to GP1 and GP2. These two fragments are linked together by disulfide-bonding (S—S). The GP1 helps in binding to the host cell and has a crucial role of EBOV entry across the host cell membrane [49]. The flowchart is shown in **Figure 4** [45].

1.7 Classification and functions of EBOV glycoproteins

The Ebola virus actually produces the following soluble glycoproteins during infection:

1.7.1 Small secreted glycoprotein of EBOV

Small secreted glycoprotein (ssGP) is translated from part of mRNA. It is nonstructural and generated when two adenosines are combined or one is deleted during transcription [50]. The ssGP has monomeric structure which has identical 295 amino acids from starting point with secreted GP and full length GP but they vary at the C-terminus of the full length glycoprotein [51]. The host ADAM17 metalloprotease enzyme is responsible to generate ssGP. The antibodies that neutralize the viral glycoproteins are quickly blocked by the secreted complex GP [14]. This complex plays a key function in the virus pathogenesis. In spite of extensive necrosis and massive virus production, it therefore contributes to less inflammatory reaction seen during the infection [14]. When the EBOV enters the host cell,

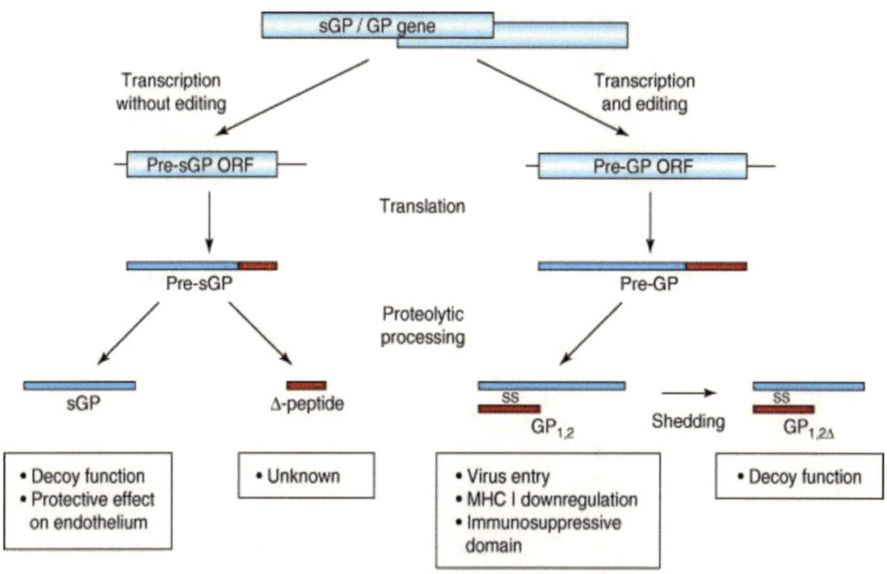

Figure 4.
Biosynthesis of glycoproteins of EBOV [45].

the small secretary glycoprotein neutralizes the host cell neutrophils through CD16b and resulting in lymphocyte death and further vascular dysregulation [52].

1.7.2 Secreted or soluble glycoprotein and delta peptide of EBOV

The soluble glycoprotein (sGP) is a non-structural protein with a single frame and has total 364-372 residues. At the N-terminal, 295 residues which also include the signal sequence are similar with full length GP. It differs in the length of C-terminal sequences [46, 50]. The formation of sGP is from Precursor (pre-sGP). Unedited mRNA is frequently produced from infected cells which results in the formation of precursor (pre-sGP). Precursor (pre-sGP) is cleaved by the furin protein to produce sGP and delta peptide [46, 50]. The orientation of sGP is antiparallel and linked together by disulfide bonding. It possesses dimerization between amino and carboxy groups as shown in **Figure 5**. The sGP plays a role in the evasion of humoral immune response by absorbing elicited antibodies. It also predicted to be involved in interaction with neutrophils of the host cells and their neutralization [53].

1.7.3 Envelope or full length glycoprotein of EBOV

It is generated on the surface of mature infectious virions and formed when RNA is edited by the process of transcription. The GP protein has 676 amino acids and is a structural polypeptide chain. It is encoded in two frames and functions in attachment and entry of virus into the defenseless cells of the host [48]. The GP is sub-divided into two glycoproteins; GP1 and GP2. They are membrane associated proteins and are linked with each other through disulfide bonding as shown in **Figure 6**. The GP1 is 130 kDa and GP2 is 24 kDa in size [47]. The GP1 helps in attachment to the receptors which are present on target cells [54]. The interactions allow the binding of viral particles to the dendritic cells and thus facilitates virus into cell entry [48]. GP2 is known as a class I viral fusion protein. Some characteristics of GP2 subunit protein of EBOV are as follows: towards the C-termini is the transmembrane helix while the "fusion- peptide" is found at the N-terminus [55]. The formation of this structure drives to combine the cell membranes of target cells and EBOV. It also helps the particles to enter into the cytoplasm of a healthy cell. The penetration of virus is mediated by the fusion of the membranes of endosomes with the endocytosed virus particle [56].

Survey of the literature showed that currently there are no crystal structures of both secreted and envelope glycoproteins available which are important for designing novel drugs that can inhibit the attachment of these glycoprotein to the host cells receptors and also to halt the activity of the secretary and soluble glycoproteins. In one of our study, we have constructed 3D structures through homology modeling [57]. The molecular dynamics simulation study of the obtained homology models and several drug docking results showed that these proteins can be targeted with small

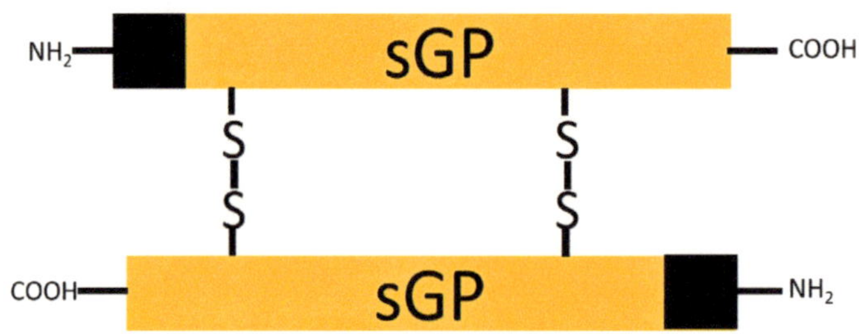

Figure 5.
EBOV soluble glycoprotein (sGP) involved in host neutrophil neutralization and immune system evasion.

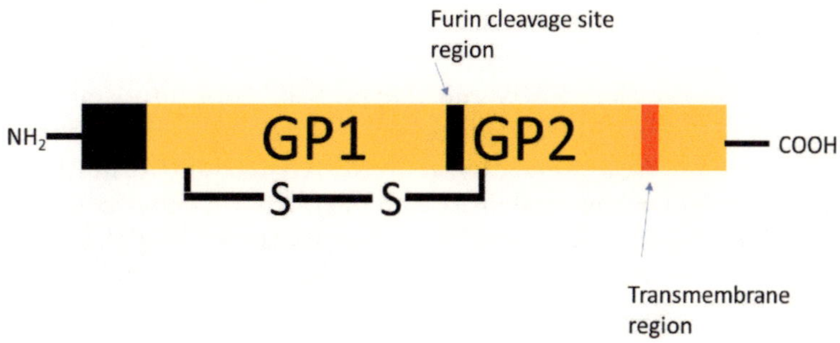

Figure 6.
EBOV envelope glycoprotein.

ligand molecules [57]. Further computation studies and experimental work will prove the usefulness of these small molecules as inhibitor of the EBOV important proteins [57]. The current advancement in cryo-electron microscopy should be utilized for the elucidation of EBOV structure like that of influenza, dengue, Zika and other arboviral diseases [57–60]. The availability of protein structure will help in molecular dynamics simulation and docking of these important viral proteins with suitable ligands and it will enhance the exploration of novel antiviral drugs [59, 60].

1.8 Progress in vaccines development for EBOV

A number of researchers have developed vaccines in their laboratories which are in initial stages and have some minor side effects [61]. A recent investigation showed the development of a vaccine when the EBOV glycoproteins like GP and VP40 were injected in the guinea pigs and a protective immune response was observed [62]. The constructed vaccine recovers the infected rhesus macaque in a single dose [62]. Similarly the glycoproteins along with the matrix proteins VP24 and VP40 and in their absence were also tested in guinea pigs [63]. The antibodies produced against these antigens completely cure the tested subjects from EBOV [63]. Several other vaccine of EBOV was developed recently and they showed promising results in murine models with high success rate [64]. Thus, a success in the development of vaccine has been achieved but passing the clinical trials and its availability in the market will take time.

1.8.1 Prevention and control

A number of steps can be taken for the control and prevention of EBOV break-out and its spread. It is always better to avoid traveling to areas of EBOV outbreak and strict quarantine should be imposed on the people of that particular towns. Only medical teams with proper safety masks, cloths should be allowed for the care of patients infected with EBOV [65]. The damaged fruits and vegetables eaten by bat should not be consumed. Bats, monkeys, pangolins, and other wild animals' meat and their utilization in food, medicine, and soaps should be avoided [66]. Contacting these animals should be discouraged. Proper washing of hands and use of sanitizers should be encouraged in schools, airports, shopping malls, trains, and all public areas. It is always necessary for hospitals and health working organizations to share and spread information related to viral diseases outbreak to the general public on time and awareness campaigns should be launched at regular intervals of time. The advancement in cryo-electron microscopy should be utilized for structural resolution of the important enzymes of the EBOV and other fatal viruses so that novel drugs can be synthesized for their inhibition and control [67].

2. Conclusions

With the 2018 outbreak of EBOV and high probability of future outbreaks and spread, it is highly important to expediate the production of effective vaccine and immunization of vulnerable population across different countries in Africa, that will help in the control of this viral disease. Joint efforts are also needed by the local public health departments and scientific community across the globe for information sharing on different viral outbreaks, vaccine development, and easy access of immunization and medicines.

Acknowledgements

We are thankful to Higher Education Commission of Pakistan for their support.

Competing interests

All the authors declared that they have no competing interests.

Note

Most part of this chapter is the introduction part of Ms. Aqsa Farman M. Phil (MS) thesis.

Author details

Aqsa Farman[1†], Syed Lal Badshah[1,2*†], Khalid Khan[1], Nasir Ahmad[1] and Abdul Naeem[2]

1 Department of Chemistry, Islamia College University, Peshawar, Pakistan

2 National Center of Excellence in Physical Chemistry, University of Peshawar, Peshawar, Pakistan

*Address all correspondence to: shahbiochemist@gmail.com

† These authors are contributed equally.

References

[1] Rajak H, Jain DK, Singh A, Sharma AK, Dixit A. Ebola virus disease: Past, present and future. Asian Pacific Journal of Tropical Biomedicine. 2015;**5**:337-343. DOI: 10.1016/S2221-1691(15)30365-8

[2] Feldmann H, Geisbert TW. Ebola haemorrhagic fever. Lancet. 2011;**377**:849-862. DOI: 10.1016/S0140-6736(10)60667-8

[3] Mühlberger E. Chapter 1. Genome organization, replication, and transcription of filoviruses. In: Klenk H-D, Feldmann H, editors. Ebola and Marburg Viruses: Molecular and Cellular Biology. Horizon Bioscience. Norfolk, UK. 2004:1-20

[4] Burk R, Bollinger L, Johnson JC, Wada J, Radoshitzky SR, Palacios G, et al. Neglected filoviruses. FEMS Microbiology Reviews. 2016;**40**: 494-519. DOI: 10.1093/femsre/fuw010

[5] Payne S. Family Filoviridae. Viruses. From Understanding to Investigation. Chapter 21; 2017. pp. 183-190. DOI: 10.1016/B978-0-12-803109-4.00021-00

[6] Sanchez A, Khan AS, Zaki SR, Nabel GJ, Ksiazek TG, Peters CJ. Filoviridae: Marburg and Ebola viruses. Fields Virology. 2001;**1**:1279-1304. DOI: 10.1007/978-1-4612-3900-0_30

[7] Peters CJ, Sanchez A, Feldmann H, Rollin PE, Nichol S, Ksiazek TG. Filoviruses as emerging pathogens. Seminars in Virology. 1994;**5**:147-154. DOI: 10.1006/smvy.1994.1015

[8] WHO, WHO Ebola Situation Reports, WHO. 2016. Available from: http://www.who.int/csr/disease/ebola/situation-reports/archive/en/

[9] Kuhn JH, Andersen KG, Bào Y, Bavari S, Becker S, Bennett RS, et al. Nichol, Filovirus refseq entries: Evaluation and selection of filovirus type variants, type sequences, and names. Viruses. 2014;**6**:3663-3682. DOI: 10.3390/v6093663

[10] Towner JS, Sealy TK, Khristova ML, Albariño CG, Conlan S, Reeder SA, et al. Newly discovered Ebola virus associated with hemorrhagic fever outbreak in Uganda. 2008;**4**: PLoS Pathogens. DOI: 10.1371/journal.ppat.1000212

[11] Gatherer D. The 2014 Ebola virus disease outbreak in West Africa. The Journal of General Virology. 2014;**95**:1619-1624. DOI: 10.1099/vir.0.067199-0

[12] Clark DV, Kibuuka H, Millard M, Wakabi S, Lukwago L, Taylor A, et al. Long-term sequelae after Ebola virus disease in Bundibugyo, Uganda: A retrospective cohort study. The Lancet Infectious Diseases. 2015;**15**:905-912. DOI: 10.1016/S1473-3099(15)70152-0

[13] Roddy P, Howard N, van Kerkhove MD, Lutwama J, Wamala J, Yoti Z, et al. Clinical manifestations and case management of Ebola haemorrhagic fever caused by a newly identified virus strain, Bundibugyo, Uganda, 2007-2008. PLoS One. 2012;7. DOI: 10.1371/journal.pone.0052986

[14] Boehmann Y, Enterlein S, Randolf A, Mhlberger E. A reconstituted replication and transcription system for Ebola virus Reston and comparison with Ebola virus Zaire. Virology. 2005;**332**:406-417. DOI: 10.1016/j.virol.2004.11.018

[15] Rollin PE, Williams RJ, Bressler DS, Pearson S, Cottingham M, Pucak G, et al. Ebola (subtype Reston) virus among quarantined nonhuman primates recently imported from the Philippines to the United States. The Journal of Infectious

Diseases. 1999;**179**:S108-S114. DOI: 10.1086/514303

[16] Kuhn JH, Becker S, Ebihara H, Geisbert TW, Johnson KM, Kawaoka Y, et al. Proposal for a revised taxonomy of the family Filoviridae: Classification, names of taxa and viruses, and virus abbreviations. Archives of Virology. 2010;**155**:2083-2103. DOI: 10.1007/s00705-010-0814-x

[17] Chiappelli F, Bakhordarian A, Thames AD, Du AM, Jan AL, Nahcivan M, et al. Ebola: Translational science considerations. Journal of Translational Medicine. 2015;**13**:11. DOI: 10.1186/s12967-014-0362-3

[18] Bitekyerezo M, Kyobutungi C, Kizza R, Mugeni J, Munyarugero E, Tirwomwe F, et al. The outbreak and control of Ebola viral haemorrhagic fever in a Ugandan medical school. Tropical Doctor. 2002;**32**:10-15. DOI: 10.1177/004947550203200107

[19] Okware SI, Omaswa FG, Zaramba S, Opio A, Lutwama JJ, Kamugisha J, et al. An outbreak of Ebola in Uganda. Tropical Medicine & International Health. 2002;**7**:1068-1075. DOI: 10.1046/j.1365-3156.2002.00944.x

[20] Green A. Uganda battles Marburg fever outbreak. Lancet. 2012;**380**:1726. DOI: 10.1016/S0140-6736(12)61973-4

[21] Pattyn SR. Ebola virus haemorrhagic fever. In: Proceedings of an International Colloquium on Ebola Virus Infection and Other Haemorrhagic Fevers held in Antwerp, Belgium, 6-8 December, 1977; 1978. xii

[22] Peters CJ, JW LD. An introduction to Ebola: The virus and the disease. National Center for infectious diseases, Centers for Disease Control and Prevention. The Journal of Infectious Diseases. 1999;**179**:ix-xvi. DOI: 10.1086/514322

[23] Busico KM, Marshall KL, Ksiazek TG, Roels TH, Fleerackers Y, Feldmann H, et al. Prevalence of IgG antibodies to Ebola virus in individuals during an Ebola outbreak, Democratic Republic of the Congo, 1995. The Journal of Infectious Diseases. 1999;**179**(Suppl):S102-S107. DOI: 10.1086/514309

[24] Le Guenno B, Formentry P, Wyers M, Gounon P, Walker F, Boesch C. Isolation and partial characterisation of a new strain of Ebola virus. Lancet. 1995;**345**:1271-1274. DOI: 10.1016/S0140-6736(95)90925-7

[25] Africa S. Outbreak of Ebola hemorrhagic fever Uganda, August 2000-January 2001. MMWR. Morbidity and Mortality Weekly Report. 2001;**50**:73-77. DOI: 10.1001/jama.285.8.1010

[26] Farrar J, Hotez PJ, Junghanss T, Kang G, Lalloo D, White NJ. Chapter 16. Viral hemorrhagic fever. In: Blumberg L, Enria D, Bausch D, editors. Manson's Tropical Diseases. 23rd ed. Elsevier. 2013:171-194.e2

[27] Olival KJ, Hayman DTS. Filoviruses in bats: Current knowledge and future directions. Viruses. 2014;**6**:1759-1788. DOI: 10.3390/v6041759

[28] Schuh AJ, Amman BR, Towner JS. Filoviruses and bats. Microbioloogy Australia. 2017:12-16. DOI: 10.1071/MA17005

[29] Pigott DM, Golding N, Mylne A, Huang Z, Henry AJ, Weiss DJ, et al. Mapping the zoonotic niche of Ebola virus disease in Africa. eLife. 2014;**3**:e04395. DOI: 10.7554/eLife.04395

[30] Leroy EM, Kumulungui B, Pourrut X, Rouquet P, Hassanin A, Yaba P, et al. Fruit bats as reservoirs of Ebola virus. Nature. 2005;**438**:575-576. DOI: 10.1038/438575a

[31] Olival KJ, Islam A, Yu M, Anthony SJ, Epstein JH, Khan SA, et al. Ebola virus antibodies in fruit bats, Bangladesh. Emerging Infectious Diseases. 2013;**19**:270-273. DOI: 10.3201/eid1902.120524

[32] Dixon MG, Schafer IJ. Ebola viral disease outbreak--West Africa, 2014. MMWR. Morbidity and Mortality Weekly Report. 2014;**63**:548-551. DOI: 10.1016/j.annemergmed.2014.10.010

[33] Nakayama E, Takada A. Ebola and Marburg viruses. Journal of Disaster Research. 2011;**6**:381-389

[34] Volchkov VE, Volchkova VA, Chepurnov AA, Blinov VM, Dolnik O, Netesov SV, et al. Characterization of the L gene and 5′ trailer region of Ebola virus. The Journal of General Virology. 1999;**80**:355-362. DOI: 10.1099/0022-1317-80-2-355

[35] Weik M, Enterlein S, Schlenz K, Muhlberger E. The Ebola virus genomic replication promoter is bipartite and follows the rule of six. Journal of Virology. 2005;**79**:10660-10671. DOI: 10.1128/JVI.79.16.10660-10671.2005

[36] Klenk H-D, Feldmann H. Ebola and Marburg viruses: Molecular and cellular biology. In: Klenk HFH-D, editor. Ebola Marbg. Molecular and Cellular Biology of Viruses. Norfolk, UK: Garland Science; 2004. p. 369

[37] Möller P, Pariente N, Klenk H-D, Becker S. Homo-Oligomerization of Marburgvirus VP35 is essential for its function in replication and transcription. Journal of Virology. 2005;**79**:14876-14886. DOI: 10.1128/JVI.79.23.14876-14886.2005

[38] Hlberger EM, Weik M, Volchkov VE, Klenk H-D, Becker S. Comparison of the transcription and replication strategies of Marburg virus and Ebola virus by using artificial replication systems. Journal of Virology. 1999;**73**:2333-2342

[39] Noda T, Sagara H, Suzuki E, Takada A, Kida H, Kawaoka Y. Ebola virus VP40 drives the formation of virus-like filamentous particles along with GP. Journal of Virology. 2002;**76**:4855-4865. DOI: 10.1128/JVI.76.10.4855-4865.2002

[40] Han Z, Boshra H, Sunyer JO, Zwiers SH, Paragas J, Harty RN. Biochemical and functional characterization of the Ebola virus VP24 protein: Implications for a role in virus assembly and budding. Journal of Virology. 2003;**77**:1793. DOI: 10.1128/JVI.77.3.1793

[41] Falzarano D, Krokhin O, Wahl-Jensen V, Seebach J, Wolf K, Schnittler HJ, et al. Structure-function analysis of the soluble glycoprotein, sGP, of ebola virus. ChemBioChem. 2006;**7**:1605-1611. DOI: 10.1002/cbic.200600223

[42] Li YH, Chen SP. Evolutionary history of Ebola virus. Epidemiology and Infection. 2014;**142**:1138-1145. DOI: 10.1017/S0950268813002215

[43] Golkar Z, Pace DG, Bagasra O. Chapter 16: Utility of potent anti-viral MicroRNAs in emerging infectious diseases. In: Biochemistry, Genetics and Molecular Biology. IntechOpen; 2016. DOI: 10.5772/61687

[44] de Wit E, Feldmann H, Munster VJ. Tackling Ebola: New insights into prophylactic and therapeutic intervention strategies. Genome Medicine. 2011;**3**. DOI: 10.1186/gm219

[45] Feldmann H, Volchkov VE, Volchkova VA, Ströher U, Klenk HD. Biosynthesis and role of filoviral glycoproteins. The Journal of General Virology. 2001;**82**:2839-2848. DOI: 10.1099/0022-1317-82-12-2839

[46] Sanchez SG, Trappier BW, Mahy CJ, Peters ST. Nichol, the virion glycoproteins of Ebola viruses are

encoded in two reading frames and are expressed through transcriptional editing. Proceedings of the National Academy of Sciences of the United States of America. 1996;**93**:3602-3607. DOI: 10.1073/pnas.93.8.3602

[47] Sanchez ZY, Yang L, Xu GJ, Crews NT, Peters CJ. Biochemical analysis of the secreted and virion glycoproteins of Ebola virus. Journal of Virology. 1998;**72**:6442-6447

[48] Lee JE, Fusco ML, Hessell AJ, Oswald WB, Burton DR, Saphire EO. Structure of the Ebola virus glycoprotein bound to an antibody from a human survivor. Nature. 2008;**454**:177-182. DOI: 10.1038/nature07082

[49] Manicassamy B, Wang J, Jiang H, Rong L. Comprehensive analysis of Ebola virus GP1 in viral entry. Journal of Virology. 2005;**79**:4793-4805. DOI: 10.1128/jvi.79.8.4793-4805.2005

[50] Volchkova VA, Feldmann H, Klenk HD, Volchkov VE. The nonstructural small glycoprotein sGP of Ebola virus is secreted as an antiparallel-orientated homodimer. Virology. 1998;**250**:408-414. DOI: 10.1006/viro.1998.9389

[51] Volchkov VE, Becker S, Volchkova VA, Ternovoj VA, Kotov AN, Netesov SV, et al. GP mRNA of Ebola virus is edited by the Ebola virus polymerase and by T7 and Vaccinia virus Polymerases1. Virology. 1995;**214**:421-430. DOI: 10.1006/viro.1995.0052

[52] Khataby K, Kasmi Y, Hammou RA, Laasri FE, Boughribi S, Ennaji MM. Ebola Virus's glycoproteins and entry mechanism. In: Ebola. IntechOpen; 2016. DOI: 10.5772/64032

[53] Towner JS, Rollin PE, Bausch DG, Sanchez A, Crary SM, Vincent M, et al. Rapid diagnosis of Ebola hemorrhagic fever by reverse transcription-PCR in an outbreak setting and assessment of patient viral load as a predictor of outcome. Journal of Virology. 2004;**78**:4330-4341. DOI: 10.1128/JVI.78.8.4330-4341.2004

[54] Sanchez A, Lukwiya M, Bausch D, Mahanty S, Sanchez AJ, Wagoner KD, et al. Analysis of human peripheral blood samples from fatal and nonfatal cases of Ebola (Sudan) hemorrhagic fever: Cellular responses, virus load, and nitric oxide levels. Journal of Virology. 2004;**78**:10370-10377. DOI: 10.1128/JVI.78.19.10370-10377.2004

[55] Volchkov VE, Blinov VM, Netesov SV. The envelope glycoprotein of Ebola virus contains an immunosuppressive-like domain similar to oncogenic retroviruses. FEBS Letters. 1992;**305**:181-184. DOI: 10.1016/0014-5793(92)80662-Z

[56] Groseth A, Ströher U, Theriault S, Feldmann H. Molecular characterization of an isolate from the 1989/90 epizootic of Ebola virus Reston among macaques imported into the United States. Virus Research. 2002;**87**:155-163. DOI: 10.1016/S0168-1702(02)00087-4

[57] Ahmad N, Farman A, Badshah SL, Ur Rahman A, Rashid HU, Khan K. Molecular modeling, simulation and docking study of ebola virus glycoprotein. Journal of Molecular Graphics & Modelling. 2017;**72**:266-271. DOI: 10.1016/j.jmgm.2016.12.010

[58] Syed S, Mabkhot YN, Naeem A, Ahmad N, Badshah SL. Zika virus, microcephaly and its possible global spread. Current Topics in Zika. 2018:1-18. DOI: 10.5772/intechopen.72507

[59] Badshah SL, Ahmad N, Ur Rehman A, Khan K, Ullah A, Alsayari A, et al. Molecular docking and simulation of Zika virus NS3 helicase. BMC

Chemistry. 2019;**13**:67. DOI: 10.1186/s13065-019-0582-y

[60] Ahmad N, Rehman AU, Badshah SL, Ullah A, Mohammad A, Khan K. Molecular dynamics simulation of zika virus NS5 RNA dependent RNA polymerase with selected novel non-nucleoside inhibitors. Journal of Molecular Structure. 2020;**1203**. DOI: 10.1016/j.molstruc.2019.127428

[61] Mire CE, Geisbert TW, Feldmann H, Marzi A. Ebola virus vaccines - reality or fiction? Expert Review of Vaccines. 2016;**15**:1421-1430. DOI: 10.1080/14760584.2016.1178068

[62] Domi A, Feldmann F, Basu R, McCurley N, Shifflett K, Emanuel J, et al. A single dose of modified Vaccinia Ankara expressing Ebola virus like particles protects nonhuman Primates from lethal Ebola virus challenge. Scientific Reports. 2018;**8**:864. DOI: 10.1038/s41598-017-19041-y

[63] Lehrer AT, Wong TAS, Lieberman MM, Johns L, Medina L, Feldmann F, et al. Recombinant subunit vaccines protect Guinea pigs from lethal Ebola virus challenge. Vaccine. 2019;**37**:6942-6950. DOI: 10.1016/j.vaccine.2019.06.035

[64] MME S, KAO M, Steffens JT, Gregory M, Vantongeren SA, Van Hoeven N, et al. Glucopyranosyl lipid adjuvant enhances immune response to Ebola virus-like particle vaccine in mice. Vaccine. 2019;**37**:3902-3910. DOI: 10.1016/j.vaccine.2019.05.026

[65] Rewar S, Mirdha D. Transmission of Ebola virus disease: An overview. Annals of Global Health. 2014:444-445. DOI: 10.1016/j.aogh.2015.02.005

[66] Dhama K, Malik YS, Malik SVS, Singh RK. Ebola from emergence to epidemic: The virus and the disease, global preparedness and perspectives. Journal of Infection in Developing Countries. 2015;**9**:441-455. DOI: 10.3855/jidc.6197

[67] Badshah S, Naeem A, Mabkhot Y. The new high resolution crystal structure of NS2B-NS3 protease of Zika virus. Viruses. 2017;**9**:7. DOI: 10.3390/v9010007

Chapter 6

Genetic Polymorphisms of Foot-and-Mouth Disease Virus

Khammadov Nail Ildarovich

Abstract

The aim of the work is to search for loci of the genome of various types of foot-and-mouth disease virus (FMDV), characterized by the lowest variability, for use as genetic markers in the polymerase chain reaction (PCR) of virus identification. The nucleotide sequences of the genomes of FMDV of types A, Asia-1, C, O, and SAT (1, 2, and 3) were analyzed. When aligning the genomes of isolates of each type of virus, potentially conservative sites were identified. Comparing these loci, different types of the virus have one, the most conserved locus. Subsequent basic local alignment search tool (BLAST) analysis established the correspondence of the conservative locus to the FMDV genome, and primers and a probe were developed to amplify this locus.

Keywords: foot-and-mouth disease virus, type, strain, genome polymorphism, PCR

1. Introduction

The foot-and-mouth disease virus (FMDV) is an RNA virus belonging to the genus *Aphthovirus,* family *Picornaviridae*. The FMDV includes serotypes: A, Asia-1, C, O, and SAT (1, 2, and 3) [1]. The FMDV infection causes ulcerations (vesicles) on the mucous membranes of the tongue, mouth, nares, hooves, and udder. FMDV is highly contagious. One sick animal can spread the infection rapidly to susceptible animals. The virus is released during the incubation period, in exhaled air, milk, urine, feces, sperm, and saliva. And when there are clinical signs of the disease, the virus is also isolated from the vesicular erosive lesions of the mucous membranes and skin [2].

FMDV is pathogenic to more than 100 domestic and wild animal species [3–5]. FMDV also affects humans [6].

The FMD is zoonotic. It can cause mild infection in humans. The virus infiltrates the human body through the mucous membranes. The incubation period (5–10 days) [7] of the disease is characterized by the virus reproduction in the primary focus. At the end of the incubation period, the pathogen spreads through the body (viraemia). In places where the virus enters, skin and mucous membrane blisters are formed. These blisters later transform into ulcers. The clinical manifestation of FMDV in children is more pronounced, and the clinical signs are sometimes similar to intoxication [8].

In addition to the obvious clinical picture, the detection of FMDV actively uses serological and molecular genetic diagnostic methods. For serological diagnostics, the enzyme immunoassay method is deservedly popular [9–13]. Polymerase chain reaction and virus genome sequencing methods are used for genetic indication of FMD [14, 15]. The historical development of FMDV research covers a wide variety of techniques. In 1960, the use of cell cultures infected with FMDV was mentioned [16]. In 1966, a complement fixation test (CFT) was used to identify the FMDV [17]. The virus neutralization was employed [18] and, in 1979, the ELISA was introduced and currently in use [9, 10, 13]. The present century is also characterized by the active use of genetic analysis of FMDV nucleic acids [19].

The aim of the study is to search for universal conservative genome loci present in all types of foot-and-mouth disease virus, for use as genetic markers for PCR virus detection.

2. Materials and methods

The methodology used in this work for analysis of the genomes of the seven FMDV serotypes (**Table 1**) were as described for other microorganisms [20, 21]. The nucleotide sequences of the desired virus were determined by searching the National Center for Biotechnology Information (NCBI) resource databases. The strain diversity of the detected FMDV using the analyzed genetic marker was determined using the nBLAST program utility, and the nucleotide sequences of the primers and probes were designed using the Vector NTI 9.1.0 program (Invitrogen Corporation).

A positive control plasmid DNA was used for PCR amplification. PCR was performed on a C1000 amplifier with a CFX96 optical unit (BioRad). The methodology for PCR amplification is similar to that described previously [20], with the following modifications: probe for PCR, direct and reverse primers were developed in the framework of this work; the primer annealing temperature was 58.5°C; and the PCR (fluorescence) result was detected on each PCR cycle, at 58.5°C via Rox and Cy5 channels.

FMDV serotype	Isolates/strains name (NCBI coding)	GenBank ID	Source
Serotype A	Isolate *Zambia/90*	MH053307	Host "cattle," country "Zambia: Kasama"
	Isolate *4235*	JN099688	Host "cattle," country "Iraq: Altenia, Hilla, Babil"
	Isolate *aphilippines iso50*	AY593793	Country "Philippines"
	Isolate *A/Pocheon/001/KOR/2010*	KC588943	Host "cattle," country "Iraq: Altenia, Hilla, Babil"
	Isolate *MAY/23/2013*	KY322678	Host "cattle," country "South Korea"
	Isolate *A01L*	KY404934	Country "Argentina"
	Isolate *A/NIG/3/15*	MG725874	Host "Bos taurus," country "Nigeria"
Serotype Asia-1	Isolate *BAN/TA/Ma-167/2013*	MF782478	Host "cattle," country "Bangladesh: Madhupur, Tangail"
	Isolate *As/SIN/PAK/L2810/2009*	JN006720	Host "cattle," country "Iraq"
	Strain *Asia-1/Jiangsu/China/2005*	EF149009	Isolation source "bovine vesicular tongue tissue lysate," country "China: Jiangsu, Wuxi"
	Isolate *IND 101-99*	DQ989310	Host "cattle," country "India"
	Isolate *IND 37-02*	DQ989311	Host "buffalo," country "India"

FMDV serotype	Isolates/strains name (NCBI coding)	GenBank ID	Source
Serotype C	Strain *C-S8p200*	FJ824812	Country "Spain"
	Isolate *c1noville iso56*	AY593804	Plum Island Animal Disease Center Virus Collection, isolated 1965, country "Switzerland"
	Isolate *cwald iso32*	AY593810	Plum Island Animal Disease Center Virus Collection, isolated 1970, country "the United Kingdom"
	Isolate *KEN/1/2004*	KM268897	Host "cattle," country "Kenya: Koibatek, Rift Valley"
	Isolate *ETH/1/71*	MH053308	Host "unknown," country "Ethiopia"
	Isolate *KEN/32/70 (K267/67)*	MH053309	Host "cattle," country "Kenya: Nanyuki, Laikipia"
	Isolate *UGA/18/70*	MH053310	Host "cattle," country "Uganda: Acholi district"
Serotype O	Isolate *UGA/3/2002*	MH053318	Host "unknown," country "Uganda: Nakasongola district"
	Strain *O/YM/YN/2000*	HQ412603	Host "Sus scrofa," country "China"
	Isolate *o11indonesia iso52*	AY593813	Plum Island Animal Disease Center Virus Collection, isolated 1962, country "Indonesia"
	Isolate *BAN/NA/Ha-156/2013*	KF985189	Host "cattle," country "Bangladesh"
Serotype SAT 1	Isolate *SAT1/NIG/1/15*	MF678823	Host "Bos taurus," country "Nigeria"
	Isolate *SAT1/NIG/2/15*	MF678824	Host "Bos taurus," country "Nigeria"
	Isolate *SAT1/NIG/3/15*	MF678825	Host "Bos taurus," country "Nigeria"
Serotype SAT 2	Isolate *EGY/9/2012*	JX014255	Country "Egypt: El-Suiz"
	Isolate *PAT/1/2012*	JX014256	Host "cattle," country "Gaza Strip: Palestinian Autonomous Territories, Rafa"
	Isolate *EGY/3/2012*	KC440884	Host "cattle," country "Egypt: Garbia Governorate"
Serotype SAT 3	Isolate *ZIM/P27/90(DSA-31)*	MH053352	Host "*Syncerus caffer*," country "Zimbabwe: Dande Safari Park"
	Triticum aestivum chromosome 2	LS480641	Organism "*Triticum aestivum*"

Table 1.
Nomenclature of isolates/strains of FMDV and other organisms described in the work.

3. Results and discussion

3.1 Epidemiology

Scientific publications on the infection of humans with the FMDV began to appear as early as 1869, 1872 [22, 23]. After some time, seven different serotypes of FMDV were identified, which have a different distribution. Some serotypes have a restricted geographical distribution, for example, Asia-1, whereas others, notably serotype O, occurred in many different regions [24].

FMDV is a very dangerous disease, and the data from the Pakistan showed that mortality due to the dominant FMDV serotype "O" was from 7.74 to 21.61% [25].

Studies in Nigeria showed a high prevalence of FMDV in the wild. Thus, the incidence of cattle ranges from 39.7 to 72.8% (in various animal breeds). Also, antibodies to FMDV were found in waterbucks, elephant, wildebeests, and other animals [26]. It is important to note that the protective properties of vaccines are effective only against infection with the same subtype [24].

3.2 Genetic analysis

The following is an analysis of the variability of the nucleotide sequences of various isolates for each of the virus serotypes (the list of isolates is shown in **Table 1**). For ease of orientation in the nucleotide sequence of the genomes of the various isolates of FMDV serotype A, the designations will be indicated relative to the isolate Zambia/90 (GenBank ID MH053307). For orientation in the nucleotide sequence of the genomes of various isolates of FMDV serotype Asia-1, designations will be indicated relative to isolate BAN/TA/Ma-167/2013 (GenBank ID MF782478). For orientation in the nucleotide sequence of the genomes of various isolates of FMDV serotype C, designations will be indicated relative to strain C-S8p200 (GenBank ID FJ824812). For orientation in the nucleotide sequence of the genomes of various isolates of FMDV serotype O, designations will be indicated relative to isolate UGA/3/2002 (GenBank ID MH053318). For orientation in the nucleotide sequence of the genomes of various isolates of FMDV SAT serotypes (1, 2, and 3), designations will be indicated with respect to SAT 3 serotype, isolate ZIM/P27/90 (DSA-31) (GenBank ID MH053352). Orientation in the nucleotide sequence of the genomes of various serotypes of FMDV, when analyzing the nucleotide sequence encoding the pathogenicity factor of the virus, will be labeled with respect to serotype A, isolate 4235 (GenBank ID JN099688). Analysis of the genomes of all serotypes is aimed at identifying the universal locus (marker nucleotide sequence), which is available in all FMDV serotypes. The marker locus should have minimal variability (minimum number of nucleotide substitutions in the annealing region of primers and probes). The size of the marker locus should not exceed 200 bp (base pair).

The identification of homologies among various types of virus, within the locus encoding the pathogenicity factor "VP1," is aimed at determining the locus, during amplification of which the maximum number of isolates (or all isolates) will be detected in all serotypes of FMDV. This nucleotide sequence is part of the nucleotide sequence encoding a viral capsid. The nucleotide sequence is located in the genome of the virus in the area from 2703 to 3353 bp. There were no significant homologies at this locus (40.5% homology), and the FMDV indication for this locus is not informative.

Further analysis was aimed at identifying homologies among the most abundant (largest number strains/isolates) serotype of the virus type, namely type O (179 strains/isolates). In the analysis, sequences were selected in which no more than two nucleotide substitutions were found in each region of the generation of oligonucleotide seeds (for different isolates). The first locus, in which it is possible to design oligonucleotide seeds, is localized in the region from 4132 to 4292 bp, while the maximum variability was observed in the strain with ID HQ412603 and isolate with ID AY593813. The second locus, in which it is possible to design oligonucleotide seeds, is localized in the region from 7810 to 7908 bp, while the maximum variability was observed in the isolate with ID KF985189; and the third locus, in which it is possible to design oligonucleotide seeds, is localized in the region from 7913 to 8043 bp; no significantly variable strains/isolates were detected in the nucleotide sequence of the locus.

Further homology was detected among the next most numerous virus strains/isolates, type A (117 strains/isolates). Homology was detected at three loci, which, based on the results of the analysis of type O virus, were able to design oligonucleotide seeds. By analyzing the first locus, which is localized in the region from 4151 to 4310 bp, one isolate with three nucleotide substitutions was identified in the region of the generation of the reverse primer (AY593793). Analysis of the second locus located in the region from 7829 to 7926 bp revealed three variable isolates (KC588943, KY322678, and KY404934). Analysis of the third locus located in the region from 7933 to 8061 bp revealed one isolate (MG725874) with three nucleotide substitutions in the region of generation of the oligonucleotide probe.

The identification of homologies among virus isolates, type Asia-1 (59 strains/isolates), was continued at the same three loci. By analyzing the first locus located in the region from 4162 to 4321 bp, one isolate (JN006720) with three nucleotide substitutions was detected in the region of generation of the oligonucleotide probe. The second locus, located in the region from 7840 to 7937 bp, was characterized by the variability of three nucleotides in the region of generation of the oligonucleotide probe in one strain (EF149009). Analysis of the third locus located in the region from 7944 to 8072 bp revealed polymorphism in the region of generation of the oligonucleotide probe in two virus isolates (DQ989310, DQ989311).

Virus isolates, type C (23 strains/isolate), were further analyzed for homology. The first locus is localized in the region from 4100 to 4259 bp, with maximum variability observed in six virus isolates (AY593804, AY593810, KM268897, MH053308, MH 053309, and MH053310). The second locus, localized in the region from 7778 to 7875 bp, with maximum variability observed in three virus isolates (AY593810, MH053308, and MH053310). And the third locus, located in the region from 7882 to 8010 bp, in the nucleotide sequence of which no significantly variable strains/isolates were detected.

The identification of homologies among virus isolates, type SAT, was carried out for a total of 74 strains/isolates for all three variants (SAT1, SAT2, and SAT3). The first locus is localized in the region from 4137 to 4296 bp, with maximum variability observed in more than 21 strain/isolates of the virus. The second locus is localized in the region from 7812 to 7909 bp, while the maximum variability was observed in 23 virus strains/isolates. And the third locus, localized in the region from 7916 to 8044 bp, with the maximum variability observed in six isolates of the virus (JX014255, JX014256, KC440884, MF678823, MF678824, and MF678825 were found to have identical nucleic substitutions, and the presented isolates had identical nucleotide sequences at the analyzed locus). An additional probe allows minimizing the effect of nucleotide substitutions on the FMDV indication (the nucleotide composition of the probe contains three substitutions characteristic of the above isolates).

Thus, the identification of homology among strains/isolates of all FMDV serotypes allowed us to determine the locus with minimal variability (in the text, this is the third locus). At this locus, oligonucleotides are complementary to the following positions in the GTA/3/2002 virus genome (GenBank ID MH053318): forward primer 7913–7934 bp, reverse primer 8026–8043 bp, and probe 7988–8024 bp.

All oligonucleotide seeds were analyzed for specificity in the nBLAST software utility. As a result of the analysis, the high specificity of the analyzed nucleotide sequences to the FMDV genome was established. In a more detailed analysis, excluding the nucleotide sequences of the FMDV genome in the search parameters, the complementarity of the forward and reverse primers was found for only one genome, *Triticum aestivum* (soft wheat), and more precisely, its second chromosome (GenBank ID LS480641.1); amplification is not possible in this case because of the large distance between the primers (527, 987, and 115 bp). The nucleotide sequences of the probes showed an identity only with the FMDV genome.

3.3 Design of oligonucleotides

In addition to specific oligonucleotide seeds, oligonucleotide seeds were designed to control the amplification reaction; in this case, cattle genes served as DNA markers, namely, the gene encoding the milk protein and the gene encoding the fat milk of cattle.

Within the above loci to designate the genome of the foot-and-mouth disease virus and DNA markers of cattle, primers and probes for PCR were designed (**Table 2**) with the following requirements: the same melting temperature of the primers (±0.5°C); minimum dimers and secondary structures; minimum GC at 3′ end; for the probe, the absence of G at the 5′ end (first nucleotide) and, most importantly, the minimum variability of the nucleotide sequence (maximum two) in the sequence of each oligonucleotide seed.

Result of the design of oligonucleotide seeds, both specific and for amplification control, is presented in **Table 2**.

Thus, the indication of all strains/isolates of all serotypes of FMDV is achieved using three modifications of the oligonucleotide probe; the first probe (P FMDV) is used to identify the main number of virus strains/isolates; the second modification of the probe (Pas FMDV) allows the detection of two isolates not detected by the previous probe viruses; serotype Asia-1, polymorphism features of nine isolates of

Detectable pathogen/marker	Primer name	Nucleotide sequence 5′⟶3′
Foot-and-mouth disease virus	Fp FMDV	atctccgtggcaggactcgc
	Rp FMDV	tgggtgaacgccgtgtgc
	P FMDV	Rox-tttgagattccaagctacagatcactttacctgc-BHQ2
	Pas FMDV	Rox-ttcgagataccaagctacagatcgctctacctgc-BHQ2
	Psat FMDV	Rox-tttgagatccctagctacagatcactttacctgc-BHQ2
The gene encoding the milk production of cattle	Fp kappa	ttggcaggcacagtatttgaca
	Rp kappa	attactaccaacagaaaccagttgca
	P kappa	Cy5-ttgaagaatttgggcaggtgacctaactg-RTQ3
The gene encoding the fat milk content of cattle DGAT	Fp DGAT	cctcttcctcaagctgttctcctac
	Rp DGAT	cctcaccagccttggcctt
	P DGAT	Cy5-acgtcaacctctggtgccgagagc-RTQ3

Table 2.
The nucleotide sequence of primers and probes for PCR.

Analyzed oligonucleotides for internal control of amplification	Fp FMDF	Rp FMDV	P FMDV	Pas FMDV	Psat FMDV
Fp kappa	—	—	—	—	—
Rp kappa	—	—	51.9%	—	51.9%
P kappa	—	—	—	—	—
Fp DGAT	—	—	—	—	—
Rp DGAT	—	—	23.2%	—	—
P DGAT	—	31.3%	—	—	—

Table 3.
Homology (%) of target oligonucleotides with test oligonucleotides to control amplification.

serotypes SAT1 and SAT2 and one isolate of serotype A, when they are indicated, the third probe is taken into account (Psat FMDV).

An analysis of the compatibility of specific and controlling PCR oligonucleotide seeds is presented in **Table 3**.

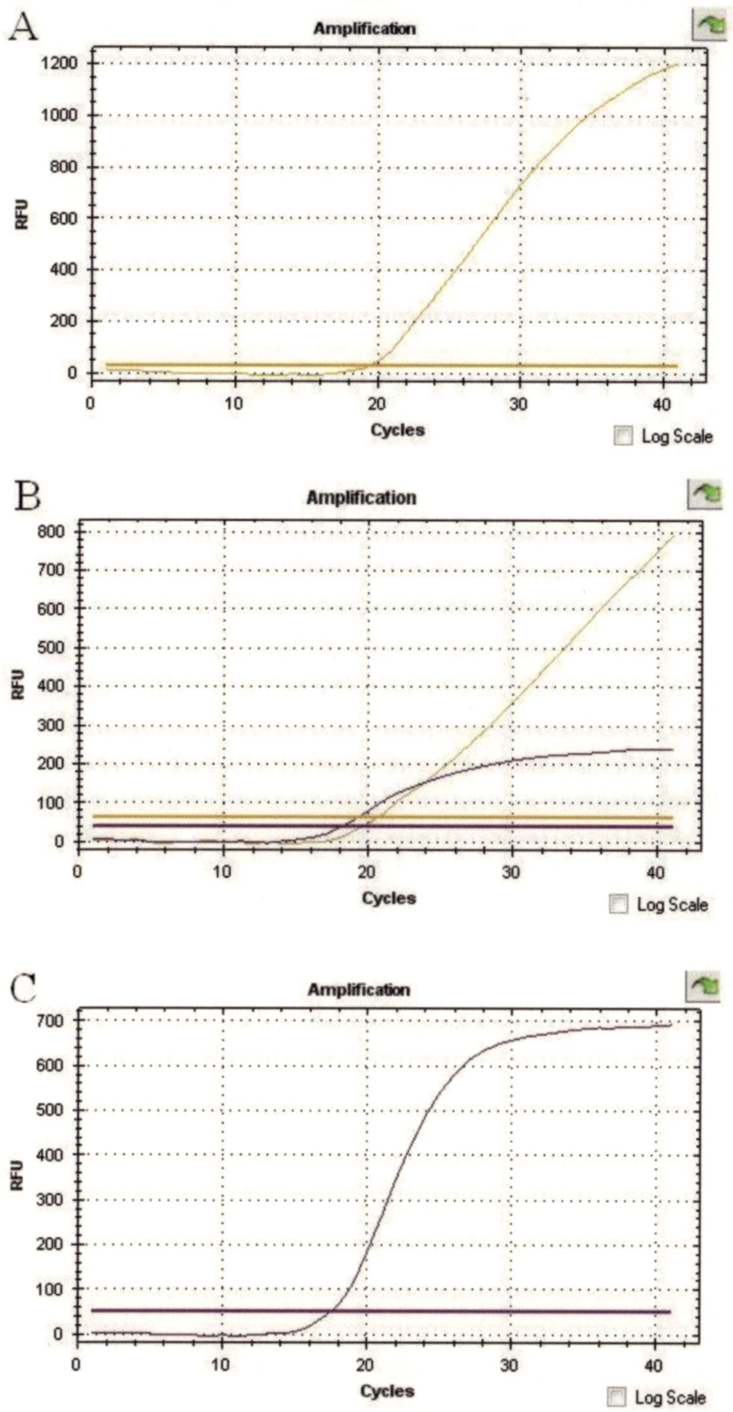

Figure 1.
Amplification of marker locus for indication of foot-and-mouth disease virus: amplification of plasmid control (A), amplification of plasmid control and cattle DNA (B), and cattle DNA amplification (C).

To control amplification, a locus was chosen within the framework of the gene encoding the milk production of cattle, since its oligonucleotide seeds are characterized by the absence of homologies with target primers to indicate FMDV, and as a result, will have minimal effect on amplification with specific primers.

The designed primers for indicating FMDV have a melting point (58.6°C ± 0.1°C), and the primers for amplification control have a melting point (55.45°C ± 0.15°C), and such a temperature difference implies more favorable amplification conditions for specific primers (at an annealing temperature of 58.5°C) and minimizes the effect of oligonucleotides to control amplification on the course of the reaction.

To control the amplification result, a positive control was created. Specific marker of positive control has the following nucleotide sequence (5′-atctccgtggcaggactcgccgtccactctggacctgacgagtaccggcgtctctttgagcccttccagggtctctttgagattccaagctacagatcactttacctgcgttgggtgaacgccgtgtgc-3′) for the indication of FMDV in plasmid DNA. The insertion of the marker sequence into the plasmid "pAL2-T" was ordered at ZAO Evrogen. To test the operability of the developed oligonucleotide seeds and plasmid control, cattle DNA and plasmid control were amplified with the primers and probes, developed above (specific for amplification control). The amplification result is shown in the **Figure 1**.

Amplification with oligonucleotide seeds for the indication of FMDV was effective both in separate PCR with positive plasmid control and in combination with cattle DNA. Amplification of the FMDV genetic markers and the control locus was performed in a single test tube. The plasmid DNA concentration was 1×10^8 DNA copies/μl (designations in **Figure 1A** and **B**). With separate amplification with cattle DNA, cross-reactions with primers/probes for indication and for foot-and-mouth disease did not occur. Amplification with oligonucleotide seeds for detecting FMDV was indicated on the Rox channel, and internal amplification was controlled on the Cy5 channel.

4. Conclusion

An analysis of the variability of the nucleotide sequences of the genomes of the different FMDV strains/isolates within each serotype revealed a locus characterized by maximum conservatism. In the sequence of oligonucleotide seeds, a sufficient level of polymorphism in the genomes of the virus isolates was found only with respect to the PCR probe (in 12 isolates by serotypes A, Asia-1, SAT1, and SAT2), and the effect of such variability on the number of detected virus isolates allows modification of the PCR probe (Pas FMDV and Psat FMDV).

Conflict of interest

The author declares no conflict of interest.

Bioethics

This work was carried out without the use of animals.

Author details

Khammadov Nail Ildarovich
Federal Center for Toxicological, Radiation and Biological Safety, Kazan City, Russian Federation

*Address all correspondence to: nikhammadov@mail.ru

References

[1] Yousef MR, Mazloum KS, Al-Nakhli HM. Serological evidence of natural exposure of camels (*Camelus dromedaries*) to foot and mouth disease virus. Veterinary World. 2012;**5**(4):197-200. DOI: 10.5455/vetworld.2012.197-200

[2] Musser JMB. A practitioner's primer on foot-and-mouth disease. Journal of the American Veterinary Medical Association. 2004;**224**(8):1261-1268. DOI: 10.2460/javma.2004.224.1261

[3] Bo LL, Lwin KS, Ungvanijban S, Knowles NJ, Wadsworth J, King DP, et al. Foot-and-mouth disease outbreaks due to an exotic serotype Asia-1 virus in Myanmar in 2017. Transboundary and Emerging Diseases. 2019;**66**(2):1067-1072. DOI: 10.1111/tbed.13112

[4] Taylor S, Khan M, Zaidi S, Alvi U, Fatima Y. Central serous retinopathy and hand–foot–mouth disease: Coincidence or causation? International Medical Case Reports Journal. 2018;**11**:277-282. DOI: 10.2147/IMCRJ.S181088

[5] Sikombe TKW, Mweene AS, Muma J, Kasanga C, Sinkala Y, Banda F, et al. Serological survey of foot-and-mouth disease virus in buffaloes (*Syncerus caffer*) in Zambia. Veterinary Medicine International. 2015:264528. DOI: 10.1155/2015/264528

[6] Miller RS, Sweeney SJ, Slootmaker C, et al. Shwiff cross-species transmission potential between wild pigs, livestock, poultry, wildlife, and humans: Implications for disease risk management in North America. Scientific Report. 2017;7:7821. DOI: 10.1038/s41598-017-07336-z

[7] Yadav S, Stenfeldt C, Branan MA, Moreno-Torres KI, Holmstrom LK, Delgado AH, et al. Parameterization of the durations of phases of foot-and-mouth disease in cattle. Frontiers in Veterinary Science. 2019;**6**:263. DOI: 10.3389/fvets.2019.00263

[8] Lizasoain A, Piegas S, Victoria M, Da Silva EE, Colina R. Hand-foot-and-mouth disease in Uruguay: Coxsackievirus A6 identified as causative of an outbreak in a rural childcare center. Journal of Medical Virology. 2020;**92**(2):167-173. DOI: 10.1002/jmv.25590

[9] Crowther JR, Abu Elzein EME. Application of the enzyme linked immunosorbent assay to the detection and identification of foot-and-mouth disease viruses. Journal of Hygiene. 1979;**83**(3):513-519. DOI: 10.1017/S002217240002636X

[10] Abu Elzein EME, Crowther JR. The specific detection of foot-and-mouth disease virus whole particle antigen (140S) by enzyme labelled immunosorbent assay. Journal of Hygiene. 1979;**83**(1):127-134. DOI: 10.1017/S0022172400025894

[11] Abu Elzein EME, Crowther JR. Serological comparison of a type sat 2 foot-and-mouth disease virus isolate from Sudan with other type SAT 2 strains. Bulletin of Animal Health and Production in Africa [Bulletin des sante et production animales en Afrique]. 1979;**27**(4):245-248

[12] Crowther JR, Abu Elzein EME. Detection of antibodies against foot-and-mouth disease virus using purified *Staphylococcus* A protein conjugated with alkaline phosphatase. Journal of Immunological Methods. 1980;**34**(3):261-267. DOI: 10.1016/0022-1759(80)90053-8

[13] Abu Elzein EME, Crowther JR. Differentiation of foot-and-mouth disease virus strains using a competition

enzyme-linked emmunosorbent assay. Journal of Virological Methods. 1982;**3**(6):355-365. DOI: 10.1016/0166-0934(82)90040-4

[14] Howson ELA, Orton RJ, Mioulet V, Lemb T, King DP, Fowler VL. Goprime: Development of an in silico framework to predict the performance of real-time PCR primers and probes using foot-and-mouth disease virus as a model. Pathogens. 2020;**9**(4):303. DOI: 10.3390/pathogens9040303

[15] Hicks HM, Wadsworth J, Azhar M, Afzal M, Manzoor S, Abubakar M, et al. Genome sequences of foot-and-mouth disease virus O/ME-SA/Ind-2001e strains isolated in Pakistan. Microbiology Resource. 2020;**9**(18):e00165-20. DOI: 10.1128/MRA.00165-20

[16] Polatnick J, Bachrach HL. Metabolic studies of bovine kidney cultures infected with foot-and-mouth disease virus. Virology. 1960;**12**(3):450-462. DOI: 10.1016/0042-6822(60)90166-5

[17] Cowan KM, Graves JH. A third antigenic component associated with foot-and-mouth disease infection. Virology. 1966;**30**(3):528-540. DOI: 10.1016/0042-6822(66)90128-0

[18] Hardy MM, Moore DM. Neutralization of foot-and-mouth disease virus. II. Further parameters related to the sensitization of the 140S virion by antibody. Journal of General Virology. 1982;**62**(2):287-295. DOI: 10.1099/0022-1317-62-2-287

[19] Klein J, Hussain M, Ahmad M, Normann P, Afzal M, Alexandersen S. Genetic characterisation of the recent foot-and-mouth disease virus subtype A/IRN/2005. Virology Journal. 2007;**4**:122. DOI: 10.1186/1743-422X-4-122

[20] Khammadov NI, Aleksandrova NM, Khammadova AV, Shuralev EA. Evaluation of the effectiveness of genetic markers of mycobacteria for assessing the disinfection quality by viability real time PCR. BioNanoScience. 2019;**9**(4):918-927. DOI: 10.1007/s12668-019-00654-8

[21] Khammadova AV, Shuralev EA, Khammadov NI, Oumarou BM, Faizov TK, Mukminov MN. Design of primers for identification of honey bee viruses in multiplex-PCR. Astra Salvensis. 2017;**1**(5):481-489

[22] Foot and mouth disease in relation to the meat and milk supply. The Lancet. 1869;**94**(2408):589-591. DOI: 10.1016/S0140-6736(02)66937-5

[23] Briscoe WT. A case of foot-and-mouth disease in the human subject. British Medical Journal. 1872;**2**(617):464. DOI: 10.1136/bmj.2.617.464

[24] Jamal SM, Belsham GJ. Foot-and-mouth disease: Past, present and future. Veterinary Research. 2013;**44**(1):116. DOI: 10.1186/1297-9716-44-116

[25] Abubakar M, Khan E, Arshed MJ, Hussain M, Ali Q, Afzal M. Mortality rate in association with foot and mouth disease outbreaks in cattle and buffaloes, Pakistan. ASM Science Journal. 2013;**7**(2):139-143 ISSN: 18236782

[26] Atuman YJ, Kudi CA, Abdu PA, Okubanjo OO, Abubakar A, Wungak Y, et al. Seroprevalence of foot and mouth disease virus infection in some wildlife and cattle in Bauchi State, Nigeria. Veterinary Medicine International. 2020:3642793. DOI: 10.1155/2020/3642793

Chapter 7

The Causative Agent of FMD Disease

Yaxin Wang and Meijun Liu

Abstract

Foot-and-mouth disease (FMD) is an acute infection of cloven-hoofed animals caused by foot-and-mouth disease virus (FMDV). It is one of the most serious infectious diseases affecting animal husbandry and a major impediment to international trade in livestock and their products. Foot-and-mouth disease virus (FMDV), a member of the *Picornaviridae* family of *Aphthovirus*, is an icosahedral virus without envelope, 25–30 nm in diameter, containing about 8.4 kb of positive-sense single-stranded RNA. The virus exists in seven different serotypes: A, O, C, Asia1, SAT1, SAT2, and SAT3, but a large number of subtypes have evolved in each serotype. This chapter reviews the genome, structure, serotype, and epidemiology of FMDV, which will help people to further explore the mechanism of the interaction between foot-and-mouth disease virus and host and provide reference for scientific prevention and control of FMDV.

Keywords: FMDV, structure, serotypes, epidemiology

1. Introduction

Foot-and-mouth disease (FMD) is an acute and highly contagious disease of cloven-hoofed animals, such as pigs, cattle, sheep, and many wild animals. Disease animals are much show fever, the place such as snout, feet, and breast forms blister and canker. The disease can spread rapidly in many ways. It has broken out many times in the world, causing huge political and economic losses to human beings. The disease was first discovered in 1514 by the Italian monk H. Fracastorius in cattle. In 1897, Loeffler and Frosch demonstrated that a filterable agent caused FMD is the foot-and-mouth disease virus (FMDV) [1]. The causative agent of FMD disease belongs to the family *Picornaviridae*, genus *Aphthovirus*. This chapter describes the status of the genome, structure, serotype, and epidemiology of the virus.

2. The genome of FMDV

The FMDV genome is a positive-sense single-stranded RNA virus with a size of about 8.5 kb [2]. FMDV RNA has a 5′ non-coding region on the left, an open reading frame (ORF) in the middle, and a 3′ non-coding region on the right. At the end of the 5′ non-coding region is a viral coding peptide VPg (or 3B), which is covalently bound to the genome. For FMDV, this 5′ untranslated region (UTR) contains S-fragment (short fragment of the genome), poly(C), pseudoknot, cre structures, and internal ribosome entry site (IRES) [3].

The S fragment can form an over 350 bases stem-ring structure, which is isolated from the genome by a variable length homopolymeric cytidylic acid tract (poly(C)), and there are some differences between the S fragments of different serotypes [4]. Carrillo et al. isolated S fragments with a sequence similarity of 80%, indicating that S fragments are highly conservative [5]. The S fragment can protect the successful replication of daughter RNA and will not be degraded by nucleic acid exonuclease, which greatly ensures the replication process of viral RNA. S fragment was involved in mediating the innate immune system. Kloc et al. found that viral RNA could not survive after deletion of more than 163 nt on stem ring of S fragment [6]. In addition, a short fragment of the G320T mutation prevented rescue of viable virus [7].

Different isolates of FMDV have different lengths of poly(C) tract; Harris and Brown found that the length of poly(C) tract may be related to the virulence of FMDV by comparing a virulent and an avirulent strain of foot-and-mouth disease virus [8]. However, other researchers suggest that the differences in virulence may be due to changes elsewhere in the genome of these strains [3].

The poly (C) tract is followed by three to four tandemly repeated pseudoknots (PKs) [9]. In a recent study, researchers compared the virulence and pathogenic mechanism of different FMDV strains in pigs and cattle by constructing PK recombinant FMDV strains and found that the absence of different sizes of PKs resulted in different pathogeny to the host, indicating that the pseudoknot region was the key to determine the viral tropism and virulence of foot-and-mouth disease virus [10].

In some picornavirus genome coding region, there is a known as cis-acting replicative element (cre) of the basic structure of RNA, cre is a conservative AAACA motif of stem loop structure, its function is to add U residues to the protein primer 3B [11]. Furthermore, Mason et al. found that cre plays an important role in genome replication and that this function is independent of its position at the 5′ end of the genome [12].

Eukaryotes generally begin translation by identifying cap structures at the 5′ end of mRNA; however, the initiation of translation can also occur internally, as has been found in picornavirus RNAs, where a functional element called the internal ribosomal entry site (IRES) at the 5′ end of mRNA also performs this function [13]. The FMDV IRES consists of 462 nucleotides with 5 domains [14]. Earlier studies have found that the interaction between IRES and the translation initiation factor eIF4G, which acts as a linker during translation initiation, is the key to in vivo translation [15]. Later, the researchers found an interaction between the IRES of FMD virus and three other translation initiation factors eIF3, eIF4B, and eIF4GII during translation initiation [16]. Furthermore, IRES trans-acting factor (ITAF) (45) promoted IRES-mediated translation in all cells; however, IRES-mediated translation activity was independent of the host range of FMDV, and only the effects of polypyrimidine tract binding protein (PTB) and eukaryotic initiation factor 4E-binding protein 1 (4E-BP1) were observed in FMDV-sensitive cells [17]. In addition, Ras GTPase SH3 domain binding protein 1 (G3BP1) interacts directly with FMDV IRES to negatively regulate translation [18].

The genome of FMDV contains an open reading frame, and ORF encodes four structural proteins (VP1, VP2, VP3, and VP4) and 10 non-structural proteins (L, 2A, 2B, 2C, 3A, 3B1, 3B2, 3B3, 3C, and 3D), whose functions will be detailed in the following sections [19, 20].

The 3′ -terminal region of FMDV consists of two distinct elements, a 90 nt untranslated region (3′-NCR) and a poly(A) tract, which have been found to stimulate IRES-driven translation [21]. Since the IRES are located at the 5′ UTR, it was assumed that there was a connection between the 3′ and 5′ ends of FMDV, and Serrano et al. subsequently demonstrated that different 3′ UTR elements are involved in the interaction between the IRES and the S region, suggesting that the 5′-3′ end of the bridge is in direct RNA-RNA contact and plays a role in RNA replication [22]. The absence of SL1 and SL2, two stem-ring structures in the 3′ non-coding region, affects viral infectivity, and the ΔSL1 mutation has been shown in pigs to be harmless to pigs, but to induce an immune response, which is important for the development of FMDV vaccines [23]. In addition, SL2 is an essential component of virus replication [23, 24].

3. Structure of the FMDV

3.1 Structural proteins

FMDV is an icosahedral symmetry non-enveloped virus. It consists of four capsid proteins VP1, VP2, VP3, and VP4, among which VP1, VP2, and VP3 are in the outermost layer of the virus, and VP4 is located in the interior of the virus and contacts with RNA [25]. Malik et al. obtained a high-resolution structure (5.2 Å) of the icosahedron of FMDV using cryo-electron microscopy (cryo-EM), notably, the obtained structure did not contain VP4 [26]. FMDV capsids are susceptible to low pH and high temperature and dissociate into pentamers under acidic conditions and release RNA [27].

In 1982, Barteling et al. proposed that FMDV structural protein (VP1) might be involved in early virus-cell interactions [28]. In the second year, Dawe and King found that the early and late viral virulence obtained by infecting BHK21 cells was different, and the researchers found that the point mutations of the VP1 were the cause of mouse virulence and BHK21 cell pathogenicity [29]. Since most of the VP1 protein is exposed on the surface of the virus, which determines the antigenicity of the virus to a large extent, VP1 protein can induce the body to produce specific neutralizing antibodies and induce anti-infection immunity [30–34]. On the VP1 of FMDV, there is a well-known G-H loop containing a highly conserved Arg-Gly-Asp (RGD) sequence, which is necessary for the virus to adhere to the cell [35, 36]. The researchers used this property of VP1 to make many explorations in the development of a vaccine against FMDV [33, 37–42]. In addition, studies have shown that VP1 N terminal is related to pH stability of FMD virus particles [43]. A recent study showed that VP1 inhibits the beta interferon signaling pathways by inhibiting IRF3 phosphorylation, dimerization, and nuclear translocation. However, the DnaJ heat shock protein family (Hsp40) member A3 (DNAJA3) can attenuate this effect [44].

For the structural protein VP2, the researchers believe that VP2 is associated with the persistence of FMDV [45]. Amino acid substitutions in the B-C loop of VP2 protein lead to antigenic differences in different types of FMDV, which indicates that VP2 is related to the antigenic diversity of FMDV [46]. In addition, amino acid substitutions in VP2 also affect the replication ability and virulence of the virus [46]. Interestingly, Vazquez-Calvo et al. found that tyrosine replacement of VP2 histidine enhanced the acid resistance of the FMDV capsid [47]. Further studies have shown that VP2 activates the EIF2S1-ATF4 pathway in cells and induces autophagy via the heat shock protein family B [small] member 1 (HSPB1) [48]. In addition, the researchers found applications for VP2 in vaccine development [49, 50] and detection of viral serotypes [51–53].

VP3 protein is the structural protein of FMDV. An amino acid deficiency of VP3 protein at position 59 of a foot-and-mouth disease virus was found in India, and the presence of this mutant increased the incidence of the epidemic [54–56]. In addition, the substitution of VP3 H142D for FMD virus can enhance the acid resistance of serotype A [57]. Furthermore, the researchers found that FMDV VP3 inhibited the IFN-beta signaling pathways [58] and the IFN-gamma signal transduction pathways [59]. Interestingly, Qi et al. found that host microRNA miR-1307 promotes the degradation of the viral structural protein VP3 through the proteasome pathway, suggesting that it may be developed for the treatment of foot-and-mouth disease [60].

Regions 20 to 35 of FMDV VP4 may be involved in inducing an immune response in T cells to recognize the T cell epitopes of MHC, a property that could be used to develop peptide vaccines [61, 62].

3.2 Nonstructural proteins of the FMDV

There are two initiation codes AUG in the ORF of FMDV, which can produce two forms of lead proteases, Lab (synthesized by the first AUG) and Lb (synthesized by the second AUG) [63]. Further studies found that the virus could still be produced in transfected cells when the first AUG was deleted, but not when the second AUG was deleted [64]. FMDV inhibits protein synthesis in host cells after infecting the host, which may be related to the cleavage of eukaryotic translation initiation factor 4GII (eIF4GII) induced by leader protease (L-pro) [65]. Further studies by Moral-Lopez et al. found that L-pro can increase the translation driven by IRES [66]. In addition, phylogenetic analysis of nucleotide sequence in L-pro region of FMD type O serum isolates from India revealed that all amino acid residues at the active cleavage site of L-pro sequence were conserved [67].

The P2 portion of FMDV is eventually processed into three mature peptides, 2A, 2B, and 2C [68]. FMDV 2A protein can cleave the site of 2A/2B, and the researchers applied this property to the field of biotechnology and successfully obtained bioactive proteins by expressing multiple proteins in cells [69–76]. It has been shown that the 2A polypeptide can cleaving the 2A/2B junction because it has a conserved c-terminal motif [D(V/I)E(S/T)NPGP], where the last P is the first residue of 2B, which is important for protein processing and virus replication [77, 78]. The researchers produced recombinant antigen of FMDV P1-2A3C in plant species, which can induce humoral immunity in guinea pigs [79]. In addition, the development of a genetically engineered vaccine against FMDV 2A may be an effective means of controlling foot-and-mouth disease [80, 81]. The study of 2B by Zhu et al. showed that, in the study of FMDV, 2B expression reduced the expression of retinoic acid-inducible gene I (RIG-I) through the interaction of residues of 2B carboxyl terminal amino acids 105–114 [82]. Further studies have shown that 2B also interacts with MDA5 and negatively regulates RLR-mediated IFN-beta induction [83]. In addition, Zhi et al. demonstrated that 2B activates NLRP3 inflammasome [84]. Further studies revealed that the non-structural protein 2B of FMDV interacts with eEF1G [85] and CypA [86] and plays a role in the process of virus infection and replication. For 2C, it was used to distinguish between infected and vaccinated animals [87–90]. The researchers identified 2C interacting proteins, including autophagy regulators Beclin1 [91], N-myc, and STAT interactor (Nmi) [92, 93], by yeast two-hybrid system and immunoprecipitation, which are helpful in understanding the mechanism of FMDV.

Similarly, the researchers were able to identify infected and vaccinated animals using non-structural protein 3A [94], which was more specific and sensitive than other non-structural proteins 3B and 3AB [95]. By means of yeast double hybridization, Gladue et al. identified that the interaction between 3A and host protein DCTN3 affected viral virulence [96]. In 2013, a study found that a partial deletion of 3A attenuated the foot-and-mouth disease virus in cattle [97], after 5 years, further research found that the deletion did not prevent subclinical infection [98]. The genome of FMDV contains three copies of the 3B protein (or VPg). In addition, 3A was found to inhibit interferon-beta signaling to evade the host immune system [99]. The study indicates that the 3B copy number is closely related to the virulence of the virus, and the virus containing a single 3B is less virulent, producing only mild disease [100]. By acting on the FMDV capsid precursor, P2-2A, 3C protease cleaved it into VPO, VP3, VP1, and 2A, and these three cleaved independently of each other [101]. Birtley et al. obtained a crystal structure with a resolution of 1.9 Å of 3C protease, which was folded like chymotrypsin and had a cys-his-asp catalytic triad [102]. It has been shown that 3C also attacks the host cytoskeleton during FMDV attack on the host [103]. Further studies showed that 3C protease could inhibit autophagy by degrading the autophagy-related protein ATG5-ATG12 [104]. The last non-structural protein is RNA-dependent RNA polymerase, 3D polymerase. Studies have shown that the synthesis of microRNA targeting 3D polymerase can effectively inhibit the replication of FMD virus in vitro [105, 106]. Therefore, 3D polymerase is one of the effective targets for the development of antiviral drugs targeting FMDV. 5D9, a 3D polymerase inhibitor, can effectively inhibit the replication of FMDV in host cells [107]. There are still many problems to be solved, and the specific function and mechanism of FMD virus non-structural proteins need to be further explored by researchers.

4. Serotypes of the FMDV

There are seven serotypes of foot-and-mouth disease virus divided into A, O, C, Asia-1, SAT 1, SAT 2, and SAT 3, and there are many subtypes of each serotype. Most of the world has had outbreaks of foot-and-mouth disease, the most common of which is serotype O. Six of the seven serotypes (A, O, C, SAT1, SAT2, and SAT3) have occurred in Africa, while four serotypes (O, A, C, Asia1) in Asia and only three serotypes (O, A, C) in South America [108]. However, there are also SAT 1 and SAT 2 viruses from as well as from Africa entering the Middle East [108]. In addition, the most recent outbreak of foot-and-mouth disease caused by serotype C virus occurred in 2004 and is now probably extinct [109].

5. Epidemiology of the FMDV

The epidemiology of FMDV includes the source of infection and the route of transmission. Foot-and-mouth disease (FMD) has the epidemiological characteristics of rapid epidemic, wide spread, and acute onset. The main source of infection is sick animals and the incubation period of animals, the incubation period 1–7 days, the average 2-4 days. Foot-and-mouth disease mainly affects artiodactyls, mainly cattle, especially calves, followed by pigs, camels, sheep, goats, and wild animals. In addition, the virus was found in blisters, milk, urine, saliva, tears, and feces of sick animals. The transmission route is extensive, which can be transmitted to susceptible animals either by direct contact or by indirect contact (e.g., secretions, feces, animal products, contaminated air, feed, etc.). Foot-and-mouth disease occurs frequently in the spring and fall. Clinical features are blister rash in the oral mucosa, hoof, and breast skin. This disease has broken out in the world several times, causing huge political and economic losses.

6. Conclusions

Foot-and-mouth disease will reduce the milk production of sick animals; severe cases will cause acute death; animal husbandry production caused a great loss, so many countries in the world to foot-and-mouth disease as the most important animal quarantine object. In the world, the United States and other developed countries have completely eliminated foot-and-mouth disease; however, in the developing countries, foot-and-mouth disease still exists. There are seven serotypes of FMD virus, which cannot be immune to each other due to their different antigens. Vaccination is a reliable and effective method for specific prevention of FMD, and a safe and effective vaccine is a prerequisite for the successful prevention, control, and eventual elimination of FMD. Therefore, in order to effectively prevent and control foot-and-mouth disease, it is necessary to thoroughly study the mechanism of action of the virus and develop more effective prevention and control methods to ensure the healthy development of animal husbandry.

Author details

Yaxin Wang* and Meijun Liu
School of Life Science, Tianjin University, Tianjin, China

*Address all correspondence to: wangyaxin@tju.edu.cn

References

[1] Brown F. The history of research in foot-and-mouth disease. Virus Research. 2003;**91**(1):3-7. DOI: 10.1016/s0168-1702(02)00268-x

[2] Domingo E, Baranowski E, Escarmis C, Sobrino F. Foot-and-mouth disease virus. Comparative Immunology, Microbiology and Infectious Diseases. 2002;**25**(5-6): 297-308. DOI: 10.1016/S0147-9571(02)00027-9

[3] Mason PW, Grubman MJ, Baxt B. Molecular basis of pathogenesis of FMDV. Virus Research. 2003;**91**(1):9-32. DOI: 10.1016/s0168-1702(02)00257-5

[4] Bunch T, Rieder E, Mason P. Sequence of the S fragment of foot-and-mouth disease virus type A12. Virus Genes. 1994;**8**(2):173-175. DOI: 10.1007/bf01703076

[5] Carrillo C, Tulman ER, Delhon G, Lu Z, Carreno A, Vagnozzi A, et al. Comparative genomics of foot-and-mouth disease virus. Journal of Virology. 2005;**79**(10):6487-6504. DOI: 10.1128/JVI.79.10.6487-6504.2005

[6] Kloc A, Segundo FDS, Schafer EA, Rai DK, Kenney M, Santos TDL, et al. Foot-and-mouth disease virus 5′-terminal S fragment is required for replication and modulation of the innate immune response in host cells. Virology. 2017;**512**:132-143. DOI: 10.1016/j.virol.2017.08.036

[7] Mohapatra JK, Pandey LK, Pattnaik B. RNA structure disrupting G320-T transversion within the short fragment of the 5′ untranslated region prevents rescue of infectious foot-and-mouth disease virus. Journal of Virological Methods. 2014;**196**:100-103. DOI: 10.1016/j.jviromet.2013.11.007

[8] Harris TJ, Brown F. Biochemical analysis of a virulent and an avirulent strain of foot-and-mouth disease virus. The Journal of General Virology. 1977;**34**(1):87-105. DOI: 10.1099/0022-1317-34-1-87

[9] Clarke BE, Brown AL, Currey KM, Newton SE, Rowlands DJ, Carroll AR. Potential secondary and tertiary structure in the genomic RNA of foot and mouth disease virus. Nucleic Acids Research. 1987;**15**(17):7067-7079. DOI: 10.1093/nar/15.17.7067

[10] Zhu ZX, Yang F, Cao WJ, Liu HN, Zhang KS, Tian H, et al. The pseudoknot region of the 5′ untranslated region is a determinant of viral tropism and virulence of foot-and-mouth disease virus. Journal of Virology. 2019;**93**(8). DOI: ARTN e02039-18/10.1128/JVI.02039-18

[11] Rieder E, Paul AV, Kim DW, van Boom JH, Wimmer E. Genetic and biochemical studies of poliovirus cis-acting replication element cre in relation to VPg uridylylation. Journal of Virology. 2000;**74**(22):10371-10380. DOI: 10.1128/jvi.74.22.10371-10380.2000

[12] Mason PW, Bezborodova SV, Henry TM. Identification and characterization of a cis-acting replication element (cre) adjacent to the internal ribosome entry site of foot-and-mouth disease virus. Journal of Virology. 2002;**76**(19):9686-9694. DOI: 10.1128/Jvi.76.19.9686-9694.2002

[13] Jackson RJ, Kaminski A. Internal initiation of translation in eukaryotes: The picornavirus paradigm and beyond. RNA. 1995;**1**(10):985-1000

[14] Martinez-Salas E, Lopez de Quinto S, Ramos R, Fernandez-Miragall O. IRES elements: Features of the RNA structure contributing to their activity. Biochimie. 2002;**84**(8):755-763. DOI: 10.1016/S0300-9084(02)01408-6

[15] Lopez de Quinto S, Martinez-Salas E. Interaction of the eIF4G initiation factor with the aphthovirus IRES is essential for internal translation initiation in vivo. RNA. 2000;**6**(10):1380-1392. DOI: 10.1017/s1355838200000753

[16] Lopez de Quinto S, Lafuente E, Martinez-Salas E. IRES interaction with translation initiation factors: Functional characterization of novel RNA contacts with eIF3, eIF4B, and eIF4GII. RNA. 2001;7(9):1213-1226. DOI: 10.1017/s1355838201010433

[17] Kanda T, Ozawa M, Tsukiyama-Kohara K. IRES-mediated translation of foot-and-mouth disease virus (FMDV) in cultured cells derived from FMDV-susceptible and -insusceptible animals. BMC Veterinary Research. 2016;**12**. DOI: 10.1186/s12917-016-0694-8

[18] Galan A, Lozano G, Pineiro D, Martinez-Salas E. G3BP1 interacts directly with the FMDV IRES and negatively regulates translation. The FEBS Journal. 2017;**284**(19):3202-3217. DOI: 10.1111/febs.14184

[19] Grubman MJ, Robertson BH, Morgan DO, Moore DM, Dowbenko D. Biochemical map of polypeptides specified by foot-and-mouth disease virus. Journal of Virology. 1984;**50**(2):579-586

[20] Feng Q, Yu H, Liu YY, He CQ, Hu JS, Sang HC, et al. Genome comparison of a novel foot-and-mouth disease virus with other FMDV strains. Biochemical and Biophysical Research Communications. 2004;**323**(1):254-263. DOI: 10.1016/j.bbrc.2004.08.086

[21] de Quinto SL, Saiz M, de la Morena D, Sobrino F, Martinez-Salas E. IRES-driven translation is stimulated separately by the FMDV 3′-NCR and poly(A) sequences. Nucleic Acids Research. 2002;**30**(20):4398-4405

[22] Serrano P, Pulido MR, Saiz M, Martinez-Salas E. The 3′ end of the foot-and-mouth disease virus genome establishes two distinct long-range RNA-RNA interactions with the 5′ end region. The Journal of General Virology. 2006;**87**:3013-3022. DOI: 10.1099/vir.0.82059-0

[23] Pulido MR, Sobrino F, Borrego B, Saiz M. Attenuated foot-and-mouth disease virus RNA carrying a deletion in the 3′ noncoding region can elicit immunity in swine. Journal of Virology. 2009;**83**(8):3475-3485. DOI: 10.1128/Jvi.01836-08

[24] Biswal JK, Subramaniam S, Ranjan R, Pattnaik B. Partial deletion of stem-loop 2 in the 3′ untranslated region of foot-and-mouth disease virus identifies a region that is dispensable for virus replication. Archives of Virology. 2016;**161**(8):2285-2290. DOI: 10.1007/s00705-016-2909-5

[25] Paprocka G. Foot-and-mouth disease virus and its molecular structure. Medycyna Weterynaryjna. 2006;**62**(7):753-756

[26] Malik N, Kotecha A, Gold S, Asfor A, Ren J, Huiskonen JT, et al. Structures of foot and mouth disease virus pentamers: Insight into capsid dissociation and unexpected pentamer reassociation. PLoS Pathogens. 2017;**13**(9). DOI: 10.1371/journal.ppat.1006607

[27] Doel TR, Baccarini PJ. Thermal stability of foot-and-mouth disease virus. Archives of Virology. 1981;**70**(1):21-32. DOI: 10.1007/bf01320790

[28] Barteling SJ, Wagenaar F, Gielkens AL. The positively charged structural virus protein (VP1) of foot-and-mouth disease virus (type O1) contains a highly basic part which may be involved in early virus-cell

interaction. The Journal of General Virology. 1982;**62**(Pt 2):357-361. DOI: 10.1099/0022-1317-62-2-357

[29] Dawe PS, King AM. Point mutations in polypeptide VP1 of foot-and-mouth disease virus affect mouse virulence and BHK21 cell pathogenicity. Archives of Virology. 1983;**76**(2):117-126. DOI: 10.1007/bf01311695

[30] Baxt B, Morgan DO, Robertson BH, Timpone CA. Epitopes on foot-and-mouth disease virus outer capsid protein VP1 involved in neutralization and cell attachment. Journal of Virology. 1984;**51**(2):298-305

[31] Robertson BH, Morgan DO, Moore DM. Location of neutralizing epitopes defined by monoclonal antibodies generated against the outer capsid polypeptide, VP1, of foot-and-mouth disease virus A12. Virus Research. 1984;**1**(6):489-500. DOI: 10.1016/0168-1702(84)90006-6

[32] Iarov AV, Gel'fanov VM, Grechaninova LA, Surovoi A, Vol'pina OM, Ivanov VT, et al. Antigenic structure of foot-and-mouth virus. IV. Synthesis and immunogenic properties of new fragments of the VP1 protein of food-and-mouth virus strain A22. Bioorganicheskaia Khimiia. 1989;**15**(9):1193-1205

[33] Volpina OM, Yarov AV, Zhmak MN, Kuprianova MA, Chepurkin AV, Toloknov AS, et al. Synthetic vaccine against foot-and-mouth disease based on a palmitoyl derivative of the VP1 protein 135-159 fragment of the A22 virus strain. Vaccine. 1996;**14**(14):1375-1380. DOI: 10.1016/s0264-410x(96)00038-2

[34] Haydon D, Lea S, Fry L, Knowles N, Samuel AR, Stuart D, et al. Characterizing sequence variation in the VP1 capsid proteins of foot and mouth disease virus (serotype 0) with respect to virion structure. Journal of Molecular Evolution. 1998;**46**(4):465-475. DOI: 10.1007/Pl00006327

[35] Fowler V, Bashiruddin JB, Belsham GJ, Stenfeldt C, Botner A, Knowles NJ, et al. Characteristics of a foot-and-mouth disease virus with a partial VP1 G-H loop deletion in experimentally infected cattle. Veterinary Microbiology. 2014;**169**(1-2):58-66. DOI: 10.1016/j.vetmic.2013.12.008

[36] Logan D, Abu-Ghazaleh R, Blakemore W, Curry S, Jackson T, King A, et al. Structure of a major immunogenic site on foot-and-mouth disease virus. Nature. 1993;**362**(6420):566-568. DOI: 10.1038/362566a0

[37] Sun M, Qian KX, Su N, Chang HY, Liu JX, Chen GF. Foot-and-mouth disease virus VP1 protein fused with cholera toxin B subunit expressed in Chlamydomonas reinhardtii chloroplast. Biotechnology Letters. 2003;**25**(13):1087-1092. DOI: 10.1023/A:1024140114505

[38] Liu Y, Hu R, Zhang S, Zhang F, Li Z, Wei X, et al. Expression of the foot-and-mouth disease virus VP1 protein using a replication-competent recombinant canine adenovirus type 2 elicits a humoral antibody response in a porcine model. Viral Immunology. 2006;**19**(2):202-209. DOI: 10.1089/vim.2006.19.202

[39] Zhuang J, You YJ, Chen B, Rao Z, Pan J. The immune responses of the fusion protein consisted of two copies of T-cell and B-cell epitopes of food-and-mouth disease virus VP1 type O and LTB and STI enterotoxins of Escherichia coli. Yi Chuan. 2006;**28**(5):557-562

[40] Fan H, Tong T, Chen H, Guo A. Immunization of DNA vaccine encoding C3d-VP1 fusion enhanced protective immune response against

foot-and-mouth disease virus. Virus Genes. 2007;**35**(2):347-357. DOI: 10.1007/s11262-007-0105-0

[41] Ren XG, Xue F, Zhu YM, Tong GZ, Wang YH, Feng JK, et al. Construction of a recombinant BHV-1 expressing the VP1 gene of foot and mouth disease virus and its immunogenicity in a rabbit model. Biotechnology Letters. 2009;**31**(8):1159-1165. DOI: 10.1007/s10529-009-9988-2

[42] Sedeh FM, Yazdanpanah S, Soleimanjahi H, Mahravani H, Shafaee K, Razavi MH, et al. Enhancement of immune responses against Iranian isolate of FMD-type O/IRN/1/2010 based on VP1 and human HSP70 genes and comparison with conventional vaccine. Acta Scientiae Veterinariae. 2014;**42**

[43] Caridi F, Vazquez-Calvo A, Sobrino F, Martn-Acebes MA. The pH stability of foot-and-mouth disease virus particles is modulated by residues located at the pentameric interface and in the N terminus of VP1. Journal of Virology. 2015;**89**(10):5633-5642. DOI: 10.1128/Jvi.03358-14

[44] Zhang W, Yang F, Zhu ZX, Yang Y, Wang ZF, Cao WJ, et al. Cellular DNAJA3, a novel VP1-interacting protein, inhibits foot-and-mouth disease virus replication by inducing lysosomal degradation of VP1 and attenuating its antagonistic role in the beta interferon signaling pathway. Journal of Virology. 2019;**93**(13). DOI: 10.1128/JVI.00588-19

[45] Horsington J, Zhang ZD. Consistent change in the B-C loop of VP2 observed in foot-and-mouth disease virus from persistently infected cattle: Implications for association with persistence. Virus Research. 2007;**125**(1):114-118. DOI: 10.1016/j.virusres.2006.12.008

[46] Xue M, Wang HW, Li W, Zhou GH, Tu YB, Yu L. Effects of amino acid substitutions in the VP2 B-C loop on antigenicity and pathogenicity of serotype Asia1 foot-and-mouth disease virus. Virology Journal. 2012;**9**. DOI: 10.1186/1743-422x-9-191

[47] Vazquez-Calvo A, Caridi F, Sobrino F, Martin-Acebes MA. An increase in acid resistance of foot-and-mouth disease virus capsid is mediated by a tyrosine replacement of the VP2 histidine previously associated with VP0 cleavage. Journal of Virology. 2014;**88**(5):3039-3042. DOI: 10.1128/Jvi.03222-13

[48] Sun P, Zhang SM, Qin XD, Chang XN, Cui XR, Li HT, et al. Foot-and-mouth disease virus capsid protein VP2 activates the cellular EIF2S1-ATF4 pathway and induces autophagy via HSPB1. Autophagy. 2018;**14**(2):336-346. DOI: 10.1080/15548627.2017.1405187

[49] Ganji VK, Biswal JK, Lalzampuia H, Basagoudanavar SH, Saravanan P, Selvan RPT, et al. Mutation in the VP2 gene of P1-2A capsid protein increases the thermostability of virus-like particles of foot-and-mouth disease virus serotype O. Applied Microbiology and Biotechnology. 2018;**102**(20):8883-8893. DOI: 10.1007/s00253-018-9278-9

[50] Biswal JK, Mohapatra JK, Bisht P, Subramaniam S, Sanyal A, Pattnaik B. A positively charged lysine residue at VP2 131 position allows for the enhanced adaptability of foot-and-mouth disease virus serotype A in BHK-21 cells. Biologicals. 2015;**43**(1):71-78. DOI: 10.1016/j.biologicals.2014.07.001

[51] Salem R, El-Kholy AA, Ibrahim M. Eight novel single chain antibody fragments recognising VP2 of foot-and-mouth disease virus serotypes A, O, and SAT 2. Virology. 2019;**533**: 145-154. DOI: 10.1016/j.virol.2019.05.012

[52] Ali W, Habib M, Shahd M. Evaluation of PCR primers targeting the VP2 region of the foot-and-mouth disease virus for improved serotype

detection. Turkish Journal of Veterinary and Animal Sciences. 2018;**42**(4):335-345. DOI: 10.3906/vet-1801-17

[53] Liu WM, Yang BL, Wang MX, Liang WF, Wang HW, Yang DC, et al. Identification of a conserved conformational epitope in the VP2 protein of foot-and-mouth disease virus. Archives of Virology. 2017;**162**(7):1877-1885. DOI: 10.1007/s00705-017-3304-6

[54] Mohapatra JK, Priyadarshini P, Pandey L, Subramaniam S, Sanyal A, Hemadri D, et al. Analysis of the leader proteinase (L-pro) region of type A foot-and-mouth disease virus with due emphasis on phylogeny and evolution of the emerging VP3(59)-deletion lineage from India. Virus Research. 2009;**141**(1):34-46. DOI: 10.1016/j.virusres.2008.12.012

[55] Mohapatra JK, Sahu A, Pandey L, Sanyal A, Hemadri D, Pattnaik B. Genetic characterization of type A foot-and-mouth disease virus 3A region in context of the reemergence of VP3(59)-deletion lineage in India. Infection, Genetics and Evolution. 2009;**9**(4):483-492. DOI: 10.1016/j.meegid.2009.01.009

[56] Mohapatra JK, Pandey LK, Sharma GK, Bank SK, Pawar SS, Palsamy R, et al. Multiplex PCR for rapid detection of serotype a foot-and-mouth disease virus variants with amino acid deletion at position 59 of the capsid protein VP3. Journal of Virological Methods. 2011;**171**(1):287-291. DOI: 10.1016/j.jviromet.2010.10.016

[57] Biswal JK, Das B, Sharma GK, Khulape SA, Pattnaik B. Role of a single amino acid substitution of VP3 H142D for increased acid resistance of foot-and-mouth disease virus serotype A. Virus Genes. 2016;**52**(2):235-243. DOI: 10.1007/s11262-016-1294-1

[58] Li D, Yang WP, Yang F, Liu HA, Zhu ZX, Lian KQ, et al. The VP3 structural protein of foot-and-mouth disease virus inhibits the IFN-beta signaling pathway. The FASEB Journal. 2016;**30**(5):1757-1766. DOI: 10.1096/fj.15-281410

[59] Li D, Wei J, Yang F, Liu HN, Zhu ZX, Cao WJ, et al. Foot-and-mouth disease virus structural protein VP3 degrades Janus kinase 1 to inhibit IFN-gamma signal transduction pathways. Cell Cycle. 2016;**15**(6):850-860. DOI: 10.1080/15384101.2016.1151584

[60] Qi LL, Wang KL, Chen HT, Liu XS, Lv JL, Hou ST, et al. Host microRNA miR-1307 suppresses foot-and-mouth disease virus replication by promoting VP3 degradation and enhancing innate immune response. Virology. 2019;**535**:162-170. DOI: 10.1016/j.virol.2019.07.009

[61] Blanco E, McCullough K, Summerfield A, Fiorini J, Andreu D, Chiva C, et al. Interspecies major histocompatibility complex-restricted Th cell epitope on foot-and-mouth disease virus capsid protein VP4. Journal of Virology. 2000;**74**(10):4902-4907. DOI: 10.1128/Jvi.74.10.4902-4907.2000

[62] Filgueira MP, Wigdorovitz A, Romera A, Zamorano P, Borca MV, Sadir AM. Detection and characterization of functional T-cell epitopes on the structural proteins VP2, VP3, and VP4 of foot and mouth disease virus O1 Campos. Virology. 2000;**271**(2):234-239. DOI: 10.1006/viro.2000.0281

[63] Robertson BH, Grubman MJ, Weddell GN, Moore DM, Welsh JD, Fischer T, et al. Nucleotide and amino acid sequence coding for polypeptides of foot-and-mouth disease virus type A12. Journal of Virology. 1985;**54**(3):651-660

[64] Cao X, Bergmann IE, Fullkrug R, Beck E. Functional analysis of the two alternative translation initiation sites of

foot-and-mouth disease virus. Journal of Virology. 1995;**69**(1):560-563

[65] Gradi A, Foeger N, Strong R, Svitkin YV, Sonenberg N, Skern T, et al. Cleavage of eukaryotic translation initiation factor 4GII within foot-and-mouth disease virus-infected cells: Identification of the L-protease cleavage site in vitro. Journal of Virology. 2004;**78**(7):3271-3278. DOI: 10.1128/Jvi.78.7.3271-3278.2004

[66] Moral-Lopez P, Alvarez E, Redondo N, Skern T, Carrasco L. L protease from foot and mouth disease virus confers eIF2-independent translation for mRNAs bearing picornavirus IRES. FEBS Letters. 2014;**588**(21):4053-4059. DOI: 10.1016/j.febslet.2014.09.030

[67] Shanmugam Y, Muthukrishnan M, Singanallur NB, Villuppanoor SA. Phylogenetic analysis of the leader proteinase (L-pro) region of Indian foot and mouth disease serotype O isolates. Veterinaria Italiana. 2015;**51**(1):31-37. DOI: 10.12834/VetIt.164.473.2

[68] Rueckert RR, Wimmer E. Systematic nomenclature of picornavirus proteins. Journal of Virology. 1984;**50**(3):957-959

[69] de Felipe P, Martin V, Cortes ML, Ryan M, Izquierdo M. Use of the 2A sequence from foot-and-mouth disease virus in the generation of retroviral vectors for gene therapy. Gene Therapy. 1999;**6**(2):198-208. DOI: 10.1038/sj.gt.3300811

[70] de Felipe P, Hughes LE, Ryan MD, Brown JD. Co-translational, intraribosomal cleavage of polypeptides by the foot-and-mouth disease virus 2A peptide. The Journal of Biological Chemistry. 2003;**278**(13):11441-11448. DOI: 10.1074/jbc.M211644200

[71] Buren S, Ortega-Villasante C, Otvos K, Samuelsson G, Bako L, Villarejo A. Use of the foot-and-mouth disease virus 2A peptide Co-expression system to study intracellular protein trafficking in Arabidopsis. PLoS One. 2012;**7**(12). DOI: 10.1371/journal.pone.0051973

[72] Sun YF, Lin Y, Zhang JH, Zheng SP, Ye YR, Liang XX, et al. Double Candida antarctica lipase B co-display on Pichia pastoris cell surface based on a self-processing foot-and-mouth disease virus 2A peptide. Applied Microbiology and Biotechnology. 2012;**96**(6):1539-1550. DOI: 10.1007/s00253-012-4264-0

[73] Liu XQ, Liu HY, Chen QJ, Yang MM, Xin HY, Bai L, et al. Construction of foot-and-mouth disease virus 2A-based bicistronic expression vector and coexpression of two genes in goat mammary epithelial cells. Animal Production Science. 2013;**53**(4):335-341. DOI: 10.1071/An12235

[74] Torres V, Barra L, Garces F, Ordenes K, Leal-Ortiz S, Garner CC, et al. A bicistronic lentiviral vector based on the 1D/2A sequence of foot-and-mouth disease virus expresses proteins stoichiometrically. Journal of Biotechnology. 2010;**146**(3):138-142. DOI: 10.1016/j.jbiotec.2010.01.017

[75] Minskaia E, Nicholson J, Ryan MD. Optimisation of the foot-and-mouth disease virus 2A co-expression system for biomedical applications. BMC Biotechnology. 2013;**13**. DOI: 10.1186/1472-6750-13-67

[76] Yen HH, Scheerlinck JPY. Biological activity of ovine IL-23 expressed using a foot-and-mouth disease virus 2A self-cleaving peptide. Cytokine. 2013;**61**(3):744-746. DOI: 10.1016/j.cyto.2013.01.014

[77] Kjaer J, Belsham GJ. Modifications to the foot-and-mouth disease virus 2A peptide: Influence on polyprotein processing and virus replication. Journal

of Virology. 2018;**92**(8). DOI: 10.1128/JVI.02218-17

[78] Kjaer J, Belsham GJ. Selection of functional 2A sequences within foot-and-mouth disease virus; requirements for the NPGP motif with a distinct codon bias. RNA. 2018;**24**(1):12-17. DOI: 10.1261/rna.063339.117

[79] Pan L, Zhang YG, Wang YL, Wang BQ, Wang WX, Fang YZ, et al. Foliar extracts from transgenic tomato plants expressing the structural polyprotein, P1-2A, and protease, 3C, from foot-and-mouth disease virus elicit a protective response in Guinea pigs. Veterinary Immunology and Immunopathology. 2008;**121**(1-2):83-90. DOI: 10.1016/j.vetimm.2007.08.010

[80] Zhan-Bo Z, Ying X, Bo H, Chang L, Huijun LJ, Kuoshi JS, et al. Immunogenicity of foot and mouth disease virus type Asia 1 protein VP1-2A fused with a multi-epitope expressed in Pichia pastoris. Journal of Animal and Veterinary Advances. 2012;**11**(9):1512-1517

[81] Kotla S, Vishanath BS, Dechamma HJ, Ganesh K, Suryanarayana VVS, Reddy GR. DNA vaccine (P1-2A-3C-pCDNA) co-administered with bovine IL-18 gives protective immune response against foot and mouth disease in cattle. Veterinary Microbiology. 2016;**193**:106-115. DOI: 10.1016/j.vetmic.2016.07.007

[82] Zhu ZX, Wang GQ, Yang F, Cao WJ, Mao RQ, Du XL, et al. Foot-and-mouth disease virus viroporin 2B antagonizes RIG-I-mediated antiviral effects by inhibition of its protein expression. Journal of Virology. 2016;**90**(24):11106-11121. DOI: 10.1128/Jvi.01310-16

[83] Li M, Xin T, Gao XT, Wu J, Wang XX, Fang LC, et al. Foot-and-mouth disease virus non-structural protein 2B negatively regulates the RLR-mediated IFN-beta induction. Biochemical and Biophysical Research Communications. 2018;**504**(1):238-244. DOI: 10.1016/j.bbrc.2018.08.161

[84] Zhi XY, Zhang Y, Sun SQ, Zhang ZH, Dong H, Luo X, et al. NLRP3 inflammasome activation by foot-and-mouth disease virus infection mainly induced by viral RNA and non-structural protein 2B. RNA Biology. 2020;**17**(3):335-349. DOI: 10.1080/15476286.2019.1700058

[85] Zhang ZW, Pan L, Ding YZ, Lv JL, Zhou P, Fang YZ, et al. eEF1G interaction with foot-and-mouth disease virus nonstructural protein 2B: Identification by yeast two-hybrid system. Microbial Pathogenesis. 2017;**112**:111-116. DOI: 10.1016/j.micpath.2017.09.039

[86] Liu HS, Xue Q, Cao WJ, Yang F, Ma LN, Liu WJ, et al. Foot-and-mouth disease virus nonstructural protein 2B interacts with cyclophilin A, modulating virus replication. The FASEB Journal. 2018;**32**(12):6706-6723. DOI: 10.1096/fj.201701351

[87] Fukai K, Morioka K, Ohashi S, Yamazoe R, Yoshida K, Sakamoto K. Differentiation of foot-and-mouth disease virus-infected pigs from vaccinated pigs using a Western blotting assay based on baculovirus-expressed nonstructural proteins 2C and 3D. The Journal of Veterinary Medical Science. 2008;**70**(12):1353-1357. DOI: 10.1292/jvms.70.1353

[88] Lu ZJ, Zhang XL, Fu YF, Cao YM, Tian MN, Sun P, et al. Expression of the major epitope regions of 2C integrated with the 3AB non-structural protein of foot-and-mouth disease virus and its potential for differentiating infected from vaccinated animals. Journal of Virological Methods. 2010;**170**(1-2):128-133. DOI: 10.1016/j.jviromet.2010.09.016

[89] Wu L, Jiang T, Lu ZJ, Yang YM, Sun P, Liang Z, et al. Development and validation of a prokaryotically expressed foot-and-mouth disease virus

nonstructural protein 2C′ 3AB-based immunochromatographic strip to differentiate between infected and vaccinated animals. Virology Journal. 2011;**8**. DOI: 10.1186/1743-422x-8-186

[90] Liu ZZ, Shao JJ, Zhao FR, Zhou GQ, Gao SD, Liu W, et al. Chemiluminescence immunoassay for the detection of antibodies against the 2C and 3ABC nonstructural proteins induced by infecting pigs with foot-and-mouth disease virus. Clinical and Vaccine Immunology. 2017;**24**(8). DOI: 10.1128/CVI.00153-17

[91] Gladue DP, O'Donnell V, Baker-Branstetter R, Holinka LG, Pacheco JM, Fernandez-Sainz I, et al. Foot-and-mouth disease virus nonstructural protein 2C interacts with Beclin1, modulating virus replication. Journal of Virology. 2012;**86**(22):12080-12090. DOI: 10.1128/Jvi.01610-12

[92] Wang JC, Wang YQ, Liu J, Ding L, Zhang QH, Li XQ, et al. A critical role of N-myc and STAT interactor (Nmi) in foot-and-mouth disease virus (FMDV) 2C-induced apoptosis. Virus Research. 2012;**170**(1-2):59-65. DOI: 10.1016/j.virusres.2012.08.018

[93] Zheng W, Li XY, Wang JC, Li XQ, Cao H, Wang YQ, et al. A critical role of interferon-induced protein IFP35 in the type I interferon response in cells induced by foot-and-mouth disease virus (FMDV) protein 2C. Archives of Virology. 2014;**159**(11):2925-2935. DOI: 10.1007/s00705-014-2147-7

[94] Kumar N, Sharma R, Kakker NK. Non-structural protein 3A for differentiation of foot-and-mouth disease infected and vaccinated animals in Haryana (India). Zoonoses and Public Health. 2007;**54**(9-10):376-382. DOI: 10.1111/j.1863-2378.2007.01075.x

[95] Yakovleva AS, Shcherbakov AV, Kanshina AV, Mudrak NS, Fomina TA. Recombinant non-structural 3A, 3B and 3AB proteins of foot-and-mouth disease virus: Use in indirect ELISA for differentiation of vaccinated and infected cattle. Molecular Biology. 2006;**40, 1**:165-171

[96] Gladue DP, O'Donnell V, Baker-Bransetter R, Pacheco JM, Holinka LG, Arzt J, et al. Interaction of foot-and-mouth disease virus nonstructural protein 3A with host protein DCTN3 is important for viral virulence in cattle. Journal of Virology. 2014;**88**(5):2737-2747. DOI: 10.1128/Jvi.03059-13

[97] Pacheco JM, Gladue DP, Holinka LG, Arzt J, Bishop E, Smoliga G, et al. A partial deletion in non-structural protein 3A can attenuate foot-and-mouth disease virus in cattle. Virology. 2013;**446**(1-2):260-267. DOI: 10.1016/j.virol.2013.08.003

[98] Stenfeldt C, Arzt J, Pacheco JM, Gladue DP, Smoliga GR, Silva EB, et al. A partial deletion within foot-and-mouth disease virus non-structural protein 3A causes clinical attenuation in cattle but does not prevent subclinical infection. Virology. 2018;**516**:115-126. DOI: 10.1016/j.virol.2018.01.008

[99] Li D, Lei CQ, Xu ZS, Yang F, Liu HN, Zhu ZX, et al. Foot-and-mouth disease virus non-structural protein 3A inhibits the interferon-beta signaling pathway. Scientific Reports. 2016;**6**. DOI: 10.1038/srep21888

[100] Pacheco JM, Henry TM, O'Donnell VK, Gregory JB, Mason PW. Role of nonstructural proteins 3A and 3B in host range and pathogenicity of foot-and-mouth disease virus. Journal of Virology. 2003;**77**(24):13017-13027. DOI: 10.1128/Jvi.77.24.13017-13027.2003

[101] Kristensen T, Newman J, Guan SH, Tuthill TJ, Belsham GJ. Cleavages at the three junctions within the foot-and-mouth disease virus capsid precursor (P1-2A) by the 3C protease are mutually

independent. Virology. 2018;**522**: 260-270. DOI: 10.1016/j.virol.2018.07.010

[102] Birtley JR, Knox SR, Jaulent AM, Brick P, Leatherbarrow RJ, Curry S. Crystal structure of foot-and-mouth disease virus 3C protease. The Journal of Biological Chemistry. 2005;**280**(12):11520-11527. DOI: 10.1074/jbc.M413254200

[103] Armer H, Moffat K, Wileman T, Belsham GJ, Jackson T, Duprex WP, et al. Foot-and-mouth disease virus, but not bovine enterovirus, targets the host cell cytoskeleton via the nonstructural protein 3C(pro). Journal of Virology. 2008;**82**(21):10556-10566. DOI: 10.1128/Jvi.00907-08

[104] Fan XX, Han SC, Yan D, Gao Y, Wei YQ, Liu XT, et al. Foot-and-mouth disease virus infection suppresses autophagy and NF-kappa B antiviral responses via degradation of ATG5-ATG12 by 3C(pro). Cell Death & Disease. 2017;**8**. DOI: 10.1038/cddis.2016.489

[105] Prabhakaran VA, John L, Nagaraj V, Vijayapillai U, Reddy GR, Joyappa DH. 3D polymerase gene of foot and mouth disease virus is a potential target for siRNA mediated inhibition of viral replication. Indian Journal of Biotechnology. 2017;**16**(3):296-303

[106] Basagoudanavar SH, Ranjitha HB, Hosamani M, Kolangath SM, Selvan RPT, Sreenivasa BP, et al. Efficient inhibition of foot-and-mouth disease virus replication in vitro by artificial microRNA targeting 3D polymerase. Acta Virologica. 2019;**63**(4):475-479. DOI: 10.4149/av_2019_407

[107] Rai DK, Schafer EA, Singh K, McIntosh MA, Sarafianos SG, Rieder E. Repeated exposure to 5D9, an inhibitor of 3D polymerase, effectively limits the replication of foot-and-mouth disease virus in host cells. Antiviral Research. 2013;**98**(3):380-385. DOI: 10.1016/j.antiviral.2013.03.022

[108] Rweyemamu M, Roeder P, Mackay D, Sumption K, Brownlie J, Leforban Y, et al. Epidemiological patterns of foot-and-mouth disease worldwide. Transboundary and Emerging Diseases. 2008;**55**(1):57-72. DOI: 10.1111/j.1865-1682.2007.01013.x

[109] Jamal SM, Belsham GJ. Molecular epidemiology, evolution and phylogeny of foot-and-mouth disease virus. Infection, Genetics and Evolution. 2018;**59**:84-98. DOI: 10.1016/j.meegid.2018.01.020

Chapter 8

Foot-and-Mouth Disease in India: Past, Present and Future Outlook - A Review

S.D. Audarya

Abstract

Foot-and-mouth disease (FMD) affects domestic livestock population of India causing heavy economic losses to the animal owners. Clinical form of the disease is readily noticed in susceptible livestock population mainly cattle, buffalo and pigs but saliently maintained in carrier animals. Foot-and-mouth disease control programme (FMDCP) is run in India by adopting series of measures from clinical diagnosis of the disease in the field, to sending clinical samples for laboratory diagnosis and till selection of vaccine candidates. Vaccines are used to cover all the susceptible livestock population. This is expected to minimise economic losses to the livestock owners due to the disease. The Government of India has been carrying out intensive FMDCP in a phase wise manner since 2003–2004 and subsequently by 2017–2018; it has covered all the districts of the country. The FMDCP is intending to vaccinate all the susceptible livestock population of the country such as cattle, buffalo, sheep, goats and pigs. That exercise was adopted to make the country free of the disease till 2025–2030. Directorate on FMD is functioning untiringly in this regard and International center on FMD has been set up to serve the South Asian Association for Regional Cooperation (SAARC) region. In the present chapter merits and shortfalls in the Foot-and-mouth disease prevention and control strategy will be discussed.

Keywords: foot-and-mouth disease, livestock, prevention and control, India

1. Introduction

Foot-and-mouth disease (FMD) is one of the most important viral diseases of large ruminants in India [1]. FMD affects mainly cattle, buffalo and pig population of the country producing severe symptoms. It can also infect sheep, goats and captive and free-range wildlife population [2–5]. The affected large ruminants exhibit high fever, excessive frothy salivation, vesicles in the mouth specially on the tongue, teats and inter-digital space and decrease in milk yield due to reduced feed intake. Apart from vesicular presence on the snout, lameness is a major feature in affected pigs. The disease is caused by the Foot-and-mouth disease virus (FMDV) classified in the genus *Aphthovirus* in the family *Picornaviridae*. It is a highly contagious viral disease transmitted mainly by close contact and through aerosols and respiratory route. It produces higher morbidity percentages in susceptible population of all ages and mortality specially in young calves due to heart affections (tiger heart). The genetic material possessed by FMDV is a ribonucleic acid (RNA). During the FMDV replication, there are chances of generation of newer progeny virus particles.

Presence of FMDV infection in India, dates back to as early as 1864 and thereafter it has been reported from many parts of the country [6]. Out of the known seven FMDV serotypes (O, A, C, Asia1, SAT1, SAT2, SAT3) of FMDV found across the globe, four serotypes viz., O, A, C, and Asia 1 were reported in livestock in India, before 1995 (World Reference Laboratory, (WRL), Pirbright). Type O was reported in 1944, type A in 1959, type C in 1955 and type Asia1 in 1951.

Probably, due to the quadrivalent vaccination against FMD and for unexplained reasons, FMDV serotype C was not recorded in India from 1995 onwards. At present only three serotypes (O, A and Asia1) of FMDV are circulating in livestock population of the country [7]. Inactivated FMDV vaccines are readily available in Indian market to prevent and control FMD [8, 9]. As diagnosis and slaughter policy cannot be practiced in India (due to ethical and socio-economical reasons), routine vaccination is the best way to achieve protective antibody response against FMD in the vaccinated animals. This chapter tries to summarise India's efforts to prevent and control FMD.

2. Livestock census and contribution of livestock in the Indian economy

For a successful implementation of animal health prevention and control programme, it requires correct and authenticated data on the susceptible livestock population. In India livestock census is conducted periodically after its first beginning in the year 1919 onwards. Recently for its 20th livestock census recording of on-site livestock heads, across 270 million households and households enterprises/non-households enterprises and institutions (660 thousand villages and 89 thousand urban wards) India adopted collection of the data by using information technology and online transmission of the data through the state National Informatics Centre (NIC). The provisional statistics of the 20th livestock census have been released for the user [10, 11].

These data sets will prove very helpful for authorities of animal health departments to devise further prevention and control strategies at the event of any new outbreak. Although provisional livestock census data only highlight about the population count regardless of its health and vaccination status, it is strongly believed that at this level of counting of the livestock, the status of vaccination for individual animal also needed to be recorded at that time and data must have been released publicly. This will not only help to know about the exact health status of the livestock population in respect to vaccination but also cross-check the number of vaccinated animals claimed by authorities [12] (**Table 1**, **Figure 1**). So, technology driven livestock census and collection of correct data on livestock will help in implementation and follow up of any livestock disease control programme in India and ultimately leading to reduction in economic impacts of the disease [13].

Internationally, India ranks first in production of milk. India produced 1,76,347.35 thousand tonnes of milk, 7,655.61 thousand tonnes of meat and 41,462.72 thousand kilogrammes of wool in the year 2017–2018 [14]. The livestock sector provided and continues to provide an additional income source to many of the farming communities in India involved in the agriculture sector. It earns foreign currency by exporting livestock products.

Kind of animal	Population (million)	Change	Susceptibility to FMD
Cattle	192.49	(+)00.80%	Highly susceptible
Goat	148.88	(+)10.10%	Susceptible
Buffalo	109.85	(+)01.10%	Highly susceptible
Sheep	074.26	(+)14.10%	Susceptible
Pigs	009.06	(−)12.03%	Highly susceptible
Mithun	000.38	(+)26.66%	Susceptible
Horses and ponies	000.34	(−)45.58%	Not susceptible
Camels	000.25	(−)37.05%	Not susceptible
Donkeys	000.12	(−)61.23%	Not susceptible
Mules	000.08	(−)57.09%	Not susceptible
Yak	000.058	(−)24.67%	Susceptible

Table 1.
Susceptible livestock population to FMD in India.

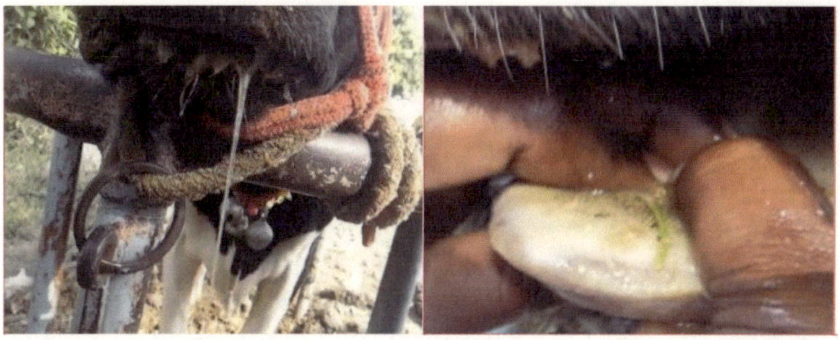

A B

Figure 1.
Cow exhibiting clinical signs of FMD. (A) Drooling of saliva; (B) Tongue lesions.

3. Literacy and its impact in the epidemiology of FMD in India

As per the population census in 2011, the literacy rate of India is 74.0% (females: 65.5% and males: 82.1%). Though gap in literacy rate is heading downwards from 21.6% in 2001 to 16.6% in 2011, there are larger variations in literacy rates in males and females among various states [15]. Illiteracy is also one of the impeding factors for personnel in livestock sector to be informed about the various activities undertaken by the central and state agencies and act upon in the event of the outbreak of the disease [16]. Though, India is implementing various educational schemes (Education to all), it is hoped that the next census will reflect improvement in literacy rates. It will help the country in respect to spreading awareness about the disease and formulation of better prevention and control strategies.

4. Foot-and-mouth disease control programme in India

Due to heavy morbidity in susceptible livestock population and negative social and economic impact of the Foot-and-mouth disease, Indian government has started progressive control pathway for FMD according to the protocols given by

Office Internationale-des-Epizooties (OIE)/Food and agriculture organisation (FAO) to minimise losses [17]. Progress for covering all the districts under the umbrella of FMDCP is presented in the **Tables 2** and **3** [18, 19].

The nine major activities that are being undertaken in the FMDCP are:

1. To vaccinate cattle and buffaloes at six monthly intervals.

2. Publicity and mass awareness campaigns.

3. Identification of target animals.

4. Sero-surveillance and monitoring of animal population on random basis.

5. Mass vaccination.

6. Procurement of cold cabinets and vaccine.

7. Quality assessment of randomly collected samples from vaccine preparations.

8. Typing of FMDV in case of outbreaks.

9. Recording/regulation of animal movement from unvaccinated areas.

In order to check whether the FMD vaccine, used in the FMDCP (at six monthly intervals), elicited protective immune response in the vaccinated large ruminants, randomly collected serum samples are being tested (**Table 2**). Paired serum samples are collected (the first serum before vaccination and the second after 21–30 days post vaccination from the same animal). These serum samples are collected from randomly selected 10 villages in each district from 10 cattle and 10 buffaloes after every vaccination. That means there are two rounds of vaccination in the year and 200 samples will be collected from each category of animals, cattle and buffaloes. So, to conclude from every district around 400 serum samples will be tested to evaluate whether the large ruminants elicited enough neutralising antibody response and protective enough to thwart future FMD outbreaks. A table indicating various institutions that are related to FMD activities is presented (**Table 3**). There are around 733 districts in India and assuming that 400 samples are generated after the end of a year (a total number of estimated serum samples screened to check for neutralising antibody response against FMDV in a year: 733 × 400 = 2,93,200). From the year 2017–2018, all the districts were covered under FMDCP, so, data for two subsequent years 2017–2018 and 2018–2019 must be made available (for a total of 5,86,400 number of serum samples) to arrive to the logical conclusions regarding

Year	Districts included	Species covered
2003–2004	54	Cattle, buffalo
2010–2011	Extended to 167 districts	
2015–2016	415	
2017–2018	All	
2019–2024	733	Cattle, buffalo, sheep, goats, pigs

Table 2.
Districts covered in FMD control programme.

Institutions	Year	Achievements and activities
Indian Veterinary Research Institute, Mukteshwar	1943	Vaccination, Virus serotyping
All India Co-ordinated Research Project, Mukteshwar	1968	For FMD virus serotyping, Central FMD laboratory at Mukteshwar and three regional centers at Hisar, Hyderabad and Kolkata
All India Co-ordinated Research Project, Mukteshwar	1971	For epidemiological studies on FMD, Facilities and manpower for extensive FMD surveillance throughout the country
Project Directorate on Foot-and-mouth disease, Mukteshwar	2001	At Mukteshwar and a network of laboratories located at various places in India, Diagnosis of FMD, Vaccine matching studies, Virus typing, Serological surveillance, Epidemiology
Directorate of Foot-and-mouth disease, Mukteshwar (http://www.pdfmd.ernet.in/)	2015	FAO reference center for FMD in South Asia, Member of a global FMD research alliance
International Center on FMD, Arugul	2017	BSL3 + Ag facility
Indian Immunologicals Limited, Hyderabad (http://www.indimmune.com/business-unit/animal-health/vaccines/livestock-vaccines), Established by National Dairy Development Board in 1982 (https://www.nddb.coop/services/rdbiotech/immunology)	1999	Vaccine manufacturing technology obtained from M/s. Wellcome foundation, United Kingdom, Has the capacity of producing 360 million trivalent doses of FMD vaccine and also exports the vaccine (Raksha Ovac, Raksha Triovac, Raksha Biovac), 80% and more vaccine for FMDCP in India
Chaudhary Charan Singh National Institute of Animal Health, Baghpat	2010	Quality control of FMD vaccine, Facilities of BSL2 and BSL3
National Institute of Veterinary Epidemiology and Disease Informatics, Bangalore (https://nivedi.res.in/about-us) formerly Project Directorate on Animal Disease Monitoring and Surveillance	2000	Epidemiology, Prediction, prevention and control of FMD threats, Weather based animal disease forecasting (National Animal Disease Referral Expert System- http://nivedi.res.in/Nadres_v2/), Use of artificial intelligence
Brilliant Biopharma Pvt. Ltd., Hyderabad (https://brilliantbiopharma.com/products/foot-and-mouth-disease-vaccine-20ml30ml100-ml/)	1988	BSL3 facility to produce animal vaccines, Supply vaccines to FAO and export vaccines (FUTVAC)
Biovet, Malur (http://biovet.in/)	2007	BSL3+ Ag production facility, Production of FMD vaccines (Bio-FMD oil), Export
Intervet India Pvt. Ltd., Pune	2004	Production of FMD vaccines, Export
Indian Agricultural Statistics Research Institute, New Delhi	2010	Center for Agricultural Bioinformatics (FMD tropism)

Table 3.
Institutes/companies that are involved in FMD research/manufacturing of FMD vaccines in India (the year represented establishment/start of functioning).

appropriate level of protective antibody response in the vaccinated animals. Early workers also commented on the weakness of the national eradication schemes in India [20]. Testing data of cattle and buffaloes from randomly selected villages as mentioned in the beginning are unavailable. Currently India is in the Stage 3 of FAO's progressive control pathway (PCP) for control of FMD [21].

The annual report of the Directorate on FMD (2017–2018) elaborated in detail on the progress of bi-annual vaccinations in different states of the country. A total of 10,02,437 serum samples were tested to assess the level of immunity. Many of the Indian states indicated development of low level of herd immunity and hence the report emphasised on regular vaccinations in the states that have low herd immunity. It was also stated that during the year 2017–2018, 21.2% cattle and buffalo and 18% sheep and goats tested positive in differentiation between infected and vaccinated animals [21]. These results when compared to the data available for the year 1995 (positivity in DIVA: 91% in cattle and buffalo and 74% in sheep and goats) indicated significant reduction in circulation of FMDV in livestock population. Thus, successful administration of FMDCP at official and ground levels and effective implementation of its mandate resulted in reduction of FMDV circulation in susceptible livestock population.

A total amount of ₹ 3.0653 billion was released by the central Government towards implementation of the control programme [12]. However, as per the data available, a total of 381.51 million livestock population (cattle and buffaloes) received the vaccine for protection against FMD in the country. Though, the state-wise data presented, for the vaccinations were not reflecting on percentages of cattle and buffalo vaccinated till date in that respective state and also for those animals yet to receive the vaccine.

5. Vaccines and vaccination for Foot-and-mouth disease in India

In India, Directorate of FMD located at Mukteshwar in the state of Uttarakhand and its collaborating and regional centers across the country are involved in continuous survey, monitoring and collection of clinical samples for virus typing and isolation from susceptible livestock population during the disease outbreaks. After isolation of a particular type of FMDV, it is stored in the virus repository for further studies. Currently, the national FMDV repository has a total of 2,188 FMDV isolates (1,482, 325, 15 and 366 belonging to the types O, A, C and Asia1 respectively) [21]. Its tally of FMDV isolates is increasing each year. These FMDV isolates are used in genotyping studies by using latest molecular biological methods and other vaccine matching experiments.

A detail on FMD vaccines is available [22]. Vaccines are one of the very important tools in the control of FMD in India [23]. Previously, India used quadrivalent inactivated FMDV vaccines (containing FMDV antigens for types O, A, C and Asia1) to control the disease. The FMDV strains used in the production of inactivated FMDV antigens for trivalent vaccines (FUTVAC) are: for type O-IND/O/R2/75, type A-IND/A/40/2000, type Asia1-IND/Asia1/63/72. The Directorate also monitors suitability of FMD antigens employed in the production of the FMD vaccines. After critical studies the directorate also recommends to the vaccine manufacturers in the country for any requirement of inclusion of suitable FMD strain of the particular FMDV type used in the vaccine production (which can confer better protection).

There were no reports in the country for the involvement of type C FMDV in the disease outbreaks hence FMDV antigen for type C was omitted from the vaccine. The absence of type C in the country may be due to implementation of vaccination programme in FMDCP or unknown reasons. At present, only trivalent inactivated FMDV vaccines are used in immunisation programmes of the livestock population in India.

India is continuously trying to increase its capability to produce sufficient doses of FMDV vaccines required in FMDCP to cover vaccination of small and large ruminants and pigs. Though, it is a big task to vaccinate susceptible livestock population but possible because of the good veterinary services [24]. Private vaccine manufacturers such as Intervet, Biovet and Brilliant Biopharma Pvt. Ltd. are ramping up the production of doses of FMD vaccines not only to cover its domestic needs but also for the export of vaccines to demanding countries. The Central Government of

India extend financial help to the Indian states for procurement and administration of vaccines to the susceptible livestock population for prevention of FMD. Present vaccines are administered at six-monthly intervals to each animal to generate sufficient level of protective immunity. Some researchers suggest that increase in the antigenic mass of the vaccine may elicit higher level of protective antibodies in the vaccinated animals (which gives protection for longer duration). Though, it seems difficult as it will increase the cost of FMD vaccines.

During FMD outbreaks, there can be economic loss of up to 80.68% attributed to reduced milk yield. Hence, to contain economic losses due to FMD, vaccination to susceptible livestock population is must. India hosts 43 indigenous cattle breeds and 13 buffalo breeds (milch and draft purpose breeds). Apart from these well documented cattle and buffalo breeds, it also holds non-descript and crossbred cattle population. Large ruminants in India are reared mainly for milk production and also for the field work in the farmland. India ranks first in the production of milk (176.3 million tonnes in 2017–2018). Milk production in India from 1950 to 1951 onwards is presented (**Figure 2**). In the country work on FMD vaccination was started from the year 1943 onwards and many vaccine manufacturers are venturing into large scale production of doses of FMDV vaccines (**Table 3**). In FMDCP, all the susceptible livestock population is being covered for conferring protection against FMD through vaccinations and thereby mitigate economic losses including that from reduced milk yield. Hence, vaccination programmes in FMDCP indirectly helped India to achieve first rank in milk production in the world.

FMD outbreaks were reported even in vaccinated animals. It leads to doubt the quality of the vaccines and vaccinations [25, 26]. It creates unnecessary fears in the livestock owners. Though, timely vaccinations reduce the incidence of the disease outbreaks and thereby reduction in antibiotic requirements for the treatment of animals for those conditions that may arise due to involvement of secondary bacterial infections during the disease outbreaks. Eventually, vaccinations help to reduce antibiotic residue in the food animals and its products [27].

Availability of rapid diagnostic testing kits for FMD allows early detection of FMD which in turn will help in devising preventative and control strategies. Early detection will help in immediate segregation of diseased animals from healthy to minimise the spread of the infection. It also alarms Government bodies to vaccinate nearby animals in the periphery from the index case (ring vaccinations) to curtail further spread [28, 29].

Critics are pointing on the fact that though vaccination is practiced in the country for prevention and control of FMD, yet, there is more than 20% of reactivity to FMD non-structural protein (NSP) 3AB3 indicative of FMDV exposure in field animals. They opined that vaccination has not resulted into generation of

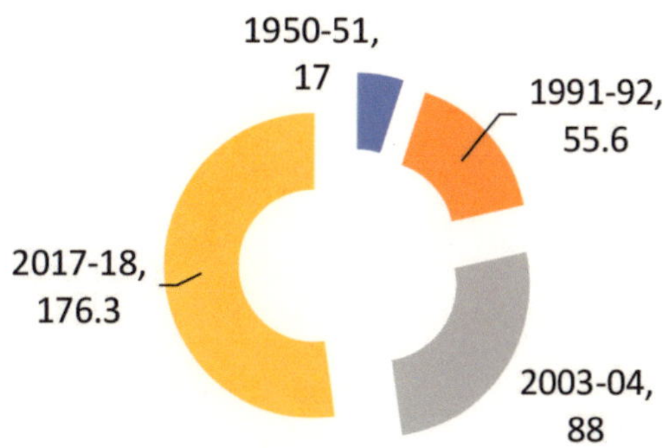

Figure 2.
Milk production in India (in million tonnes).

sufficient levels of protective immunity in the animals and poor vaccine quality. To check FMDV vaccine quality, Government of India has set up a vaccine quality testing facility at Baghpat (**Table 3**). FMDV has RNA as a genomic material. During the process of viral replication, there are chances of mutations in the FMDV genome and sometimes it may be possible that the antibodies generated after the vaccination may not be able to neutralise the circulating strain of the FMDV. Apart from this, there are many factors like persistent FMDV infections, nutritional status of the host, parasitic infestation of the host and other miscellaneous complexities prevalent in field conditions of such a vast country which can have its impact on the vaccination programme [30]. Still, FMDCP in India achieved a success in terms of reduction in FMD outbreaks as discussed below.

6. Reduction in Foot-and-mouth disease outbreaks reported in India

Foot-and-mouth disease control programme (FMDCP) is led by Department of Animal Husbandry Dairying and Fisheries, Government of India. Due to continuous efforts to prevent and control the FMDV infection in susceptible livestock population especially large ruminants, there is significant reduction in number of FMD outbreaks reported during the years 2012–2018 to that of the FMD outbreaks reported during the years 2002–2012 (**Figure 3**). Reduction in the number of outbreaks reported is largely due to comprehensive sero-monitoring, epidemiological investigations, increased diagnostic capabilities, trained manpower, competency in vaccine manufacturing capacity to cover the livestock population in the vaccination programme and Government push.

In 2019, fully central Government backed National Animal Disease Control Programme (NADCP) of ₹ 126.52 billion was launched to vaccinate 600 million animals to control FMD and Brucellosis. It is hoped that with this continuous efforts and interest of the authorities and guidelines formulated, there will be implementation of the FMDCP more effectively and efficiently [31, 32] resulting into further reduction in the incidence of the disease and ultimately eradication.

However, there must be positive criticism in few regards. As the data stated that there is no incidence of FMD in Madhya Pradesh during the year 2017–2018 but the adjoining states to Madhya Pradesh, Rajasthan, Gujarat and Uttar Pradesh reporting

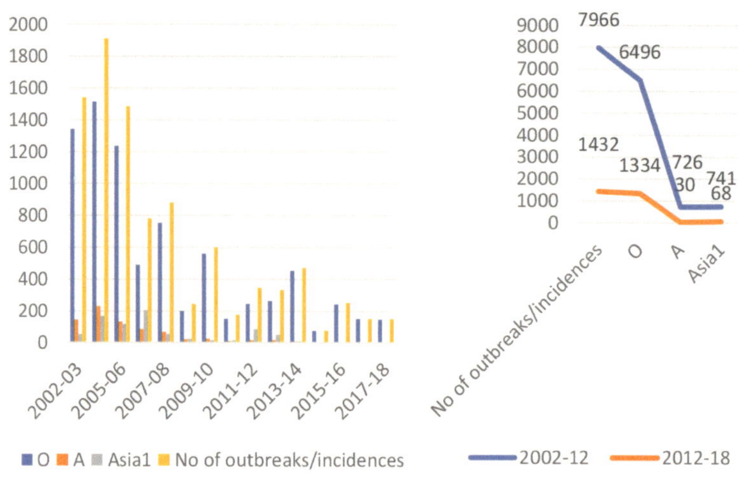

Figure 3.
FMD outbreaks/incidences in India.

the disease? Migratory small ruminants and camels are used to reach Madhya Pradesh from Rajasthan. Cattle are being procured from Gujarat, yet astonishingly the state is not reporting any incidence of FMD for the last few years. Ironically, it reported about 18% reactivity to non-structural proteins (NSP) and as a corollary more is the presence of anti-NSP antibodies, more will be the incidence of FMD. That's not holding true in this case. It can be concluded that the FMD incidence is underreported in this case. The author is a witness to a farm visit, where the cattle owner was not aware even about the vaccination practices in his last 30 years of livestock rearing. Similarly, in one of the other studies FMD was reported at Bulandshahr in Uttar Pradesh where mortalities had happened even in vaccinated animals but at a lower level than the unvaccinated ones in the year 2015 [33].

As there will be reduction in the number of outbreaks reported in a state various known and unknown pressures would play a role in hiding any further positive cases of FMD. Few random visits of authorities from a different state can be planned during vaccinations and further monitoring. Hence availability of a dedicated contact number as envisioned in NADCP for reporting the outbreak by common man or livestock owner will bring more transparency. Since each one of the eligible cattle, buffalo, sheep, goats and pigs will receive vaccine for protection against FMD, the processes of vaccinations, blood collections and follow up needed to be fully recorded digitally. The internet data is cheapest in India, presently. So, location tracing applications can be developed and used to monitor actions of vaccinators and other workers. In India, declaration of monetary incentives for reporting about the incidence of the disease will not only bring more transparency in reporting the disease but also serve a great cause to prevent and control it.

If an outbreak of FMD is noticed at a place, it needed to be reported immediately as per the FMDCP. But at the field level due to unawareness of significance of reporting an outbreak and unexplained pressures experienced by the workers, sometimes, at some places, it may go unreported. Hence it is desirable to spread awareness among common man and livestock owners regarding the programme. This programme is yet to popularize, countrywide, in livestock owners in respect to the basic understanding of the disease and its prevention and control. As there are still few pockets of the farmers left who are not willing to vaccinate their animals [33]. Such farmers needed to be identified and must be penalised else all the good of the control programme would be in danger. More coverage by the print and press media is needed to spread mass awareness. Once you start any programme subsequently it will be closely watched by the benefactor and also common public for the progressive outcome. Moreover, recently FMD outbreaks even in the vaccinated animals population raise unwarranted doubts in the animal owners unless they are better addressed about the epidemiology and transmission of the disease in a better way. Earlier states may also be reluctant to disclose the incidence of the disease due to further compliance to the provisions of the Prevention and Control of Infectious and Contagious Diseases in Animals Act, 2009. Awareness about the disease in a household is very important. A household's health-seeking behaviour influences selection into preventative care interventions [34].

7. Discussion

India is having a huge experience in recording of the data of its citizens and generating identity cards. Sometimes these cards are essential to have access to and avail certain facilities by its citizens. National level coding of individual designated livestock (if not for all) where efforts to create disease free zones are on can be implemented. Migratory flocks of small ruminants and of other animals needed to be identified and given a free pass to trace their journey. Mithun and Yaks are also susceptible for FMD but they have been kept aside from the control programme. Services of State veterinary colleges needed to be utilised in a better way and funded to for the set up of extension camps and activities which spread awareness among livestock owners. However, enthusiasm shown by veterinarians to spread awareness in the livestock owners about free FMDV vaccinations from remote locations of India in Reasi, Jammu is widely appreciated (the veterinary official took para-gliding to spread FMD awareness in the mountainous region!). There are very few reports available on impact of the FMDCP in the control of FMD [35]. Haryana state is also going to undertake trials of combined vaccines of FMD and Haemorrhagic septicaemia (HS). These combined vaccines are shown to elicit better immunity. Then the question to NADCP will be why they have chosen only Haryana for testing combined vaccines and not chosen entire country? It will be wise to go with combined vaccines for FMD and HS in the NADCP, since it will reduce the cost of vaccination and handling of the animals, save manpower, brings in additional expert manpower involved in HS research and increase its outreach. International community is closely watching the scenario as there was a report indicating transboundry movement of FMD to Sri Lanka from India [36]. So, FMD vaccines needed to be of superior quality which can confer higher level of protection by eliciting neutralising antibodies. The inactivated FMDV vaccines used in FMDCP in India, require cold chain maintenance from its production to administration in the animals. Sometimes vaccine failures can be due to improper storage of the vaccines. Presently, there is no way to confirm whether the cold chain is maintained during transportation of the vaccines or not. Vaccines can be incorporated with certain indicators whose colour may change irreversibly when exposed to higher temperatures. Secrecy is being maintained to disclose vaccine quality testing data and only few authorities are designated for the testing. Rather processes of vaccine quality testing needed to be digitally recorded and must have real time access to any viewers for positive criticism. To clear any doubts in the minds of the public due to recent controversies in vaccine manufacturing and quality control, apart from the designated national agencies for quality testing of the FMD vaccines, services of any third party national agency and if need be, international agency must be hired wherever applicable. There needed to be more coordination among institutions to share real time data for the public. Such reporting and real time sharing of the data needed to be encouraged and talked upon to relish any success stories to inspire for and to learn any lessons otherwise. Newer and newer vistas must be explored [37]. However practices based on indigenous knowledge for control and treatment of FMD from rural India is well documented [38].

8. Conclusions

India is having a will to prevent and control Foot-and-mouth disease, an economically important viral disease of livestock which causes huge annual losses of about ₹ 200 billion. In the past, India has successfully eradicated Rinderpest and Poliovirus, during those times to till now, it has tremendously enriched its know-how and institutional capabilities to handle a massive scale programme as that of FMDCP. There is a huge drop down in the number of FMD incidences/outbreaks reported as a result of this programme. India has a mechanism and expertise in place to prevent and control FMD and eradicate it by 2025–2030, but that has to be backed by the livestock owners, scientific communities, institutions and its people, enlightened for the disease by the disease, FMD.

DOI: http://dx.doi.org/10.5772/intechopen.93348

Acknowledgements

No funds are made available or received. Author wishes to extend thanks to Mr. Abhay Pratap, UG student (B.V.Sc. & A.H.) for his help.

Conflict of interest

The author declares no conflict of interest.

Author details

S.D. Audarya
Department of Veterinary Microbiology, College of Veterinary Science and Animal Husbandry, Nanaji Deshmukh Veterinary Science University, Dr. Ambedkar Nagar-Mhow, Indore, Madhya Pradesh, India

*Address all correspondence to: asd_vet@yahoo.com; audarya.sachin@gmail.com

References

[1] Department of Animal Husbandry Dairying and Fisheries Annual report [Internet]. 2018-2019. Available from: http://dadf.gov.in/sites/default/filess/Annual%20Report.pdf [Accessed: 22 February 2020]

[2] Uppal PK. Foot-and-Mouth disease in small ruminants – an issue of concern [Internet]. Available from: http://www.fao.org/ag/againfo/commissions/docs/greece04/app29.pdf [Accessed: 29 February 2020]

[3] Rout M, Senapati MR, Dash BB, Sanyal A, Pattnaik B. Serosurveillance of foot-and-mouth disease in sheep and goat population of India. Preventive Veterinary Medicine. 2014;**113**:273-277

[4] Thomson GR, Vosloo W, Bastos ADS. Foot-and-mouth disease in wildlife. Virus Research. 2003;**91**:145-161

[5] Rout M, Subramaniam S, Dash B, Mohapatra JK, Dash BB, Sanyal A, et al. Foot-and-mouth disease in wildlife population of India. Indian Journal Of Animal Research. 2017;**51**:344-346

[6] Pattnaik B, Subramaniam S, Sanyal A, Mohapatra JK, Dash BB, Ranjan R, et al. Foot-and-mouth disease: Global status and future road map for control and prevention in India. Agricultural Research. 2012;**1**:132-147. DOI: 10.1007/s40003-012-0012-z

[7] Anonymous India [Internet]. Available from: https://www.wrlfmd.org/southern-asia/india [Accessed: 21 February 2020]

[8] Srinivasan VA. FMD vaccines and vaccination in India; production, use and quality [Internet]. Available from: http://www.fao.org/docs/eims/upload/299807/an348e00.pdf [Accessed: 23 February 2020]

[9] Singh RK, Sharma GK, Mahajan S, Dhama K, Basagoudanavar SH, Hosamani M, et al. Foot and mouth disease virus: Immunobiology, advances in vaccines, and vaccination strategies addressing vaccine failures-an Indian perspective. Vaccine. 2019;**7**:1-28. DOI: 10.3390/vaccines7030090

[10] Anonymous Key statistics provisional [Internet]. 2019. Available from: http://164.100.117.97/WriteReadData/userfiles/key%20results.pdf [Accessed: 21 February 2020]

[11] Anonymous 20th Livestock census [Internet]. 2019. Available from: https://pib.gov.in/PressReleasePage.aspx?PRID=1588304 [Accessed: 21 February 2020]

[12] Anonymous Foot-and-mouth disease control programme [Internet]. Available from: http://dadfonline.gov.in/FMDCP.aspx [Accessed: 21 February 2020]

[13] Knight-Jones TJD, Rushton J. The economic impacts of foot and mouth disease – What are they, how big are they and where do they occur? Preventive Veterinary Medicine. 2013;**112**:161-173

[14] Anonymous [Internet]. Available from: http://dadfonline.gov.in/ [Accessed: 21 February 2020]

[15] Anonymous Status of literacy [Internet]. 2011. Available from: http://censusindia.gov.in/2011-prov-results/data_files/mp/07Literacy.pdf [Accessed: 07 February 2020]

[16] Anonymous National animal disease reporting system [Internet]. Available from: http://dadfonline.gov.in/NADRS.aspx# [Accessed: 21 February 2020]

[17] Anonymous Foot-and-mouth disease control programme [Internet]. Available from: http://dadfonline.gov.

in/FMDCP/Index.aspx [Accessed: 21 February 2020]

[18] Anonymous Districts of India [Internet]. 2020. Available from: https://www.mapsofindia.com/districts-india/ [Accessed: 22 February 2020]

[19] Anonymous Foot-and-mouth disease control programme [Internet]. 2019. Available from: http://dadf.gov.in/division/more-about-fmdcp1 [Accessed: 04 January 2020]

[20] Paton DJ, Sumption KJ, Charleston B. Options for control of foot-and-mouth disease: Knowledge, capability and policy. Philosophical Transactions of the Royal Society B. 2009;**364**:2657-2667. DOI: 10.1098/rstb.2009.0100

[21] Indian Council of Agricultural Research, Directorate on Foot-and-Mouth Disease [Internet]. Available from: http://www.pdfmd.ernet.in/index_files/Content/Reports/Annual_Report_2017_18.pdf [Accessed: 01 March 2020]

[22] Segundo FD, Medina GN, Stenfeldt C, Arzt J, Santos TDL. Foot-and-mouth disease vaccines. Veterinary Microbiology. 2017;**206**:102-112

[23] Saminathan M, Rana R, Ramakrishnan MA, Karthik K, Malik YS, Dhama K. Prevalence, diagnosis, management and control of important diseases of ruminants with special reference to Indian scenario. Journal of Experimental Biology and Agricultural Sciences. 2016;**4**(1):338-367. DOI: 10.18006/2016.4(3S).338.367

[24] Weaver J, Batho H, Munstermann S, Woodford J. OIE PVS Evaluation Mission Report [Internet]. 2019. Available from: https://www.oie.int/fileadmin/Home/eng/Support_to_OIE_Members/docs/pdf/25022019_India_PVS_Evaluation_report_final.pdf [Accessed: 12 February 2020]

[25] Anonymous Government to issue notice to 3 cos making FMD vaccines [Internet]. 2020. Available from: https://economictimes.indiatimes.com/industry/healthcare/biotech/pharmaceuticals/government-to-issue-notice-to-3-cos-making-fmd-vaccines/articleshow/45101376.cms?from=mdr[Accessed: 02 March 2020]

[26] Mathew L, Menon DG. Economic impact of FMD in Chazhoor Panchayath. Veterinary World. 2008;**1**:5-6

[27] Hoelzer K, Bielke L, Blake DP, Cox E, Cutting SM, Devrienndt B, et al. Vaccines as alternatives to antibiotics for food producing animals. Part-1 challenges and needs. Veterinary Research. 2018;**49**:64

[28] Sharma GK, Mahajan S, Matura R, Subramaniam S, Ranjan R, Biswas J, et al. Diagnostic assays developed for the control of foot-and-mouth disease in India. World Journal of Virology. 2015;**4**:295-302. DOI: 10.5501/wjv.v4.i3/295

[29] Singh A, Kumar M, Verma AK, Nirwan S. Studies on diagnosis of foot and mouth disease by ELISA and reverse transcription polymerase chain reaction in bovines. International Journal of Livestock Research. 2019;**9**:130-135. DOI: 10.5455/ijlr.20190325112442

[30] Artz J, Belsham GL, Lohse L, Botner A, Stenfeldt C. Transmission of foot-and-mouth disease from persistently infected carrier cattle to naïve cattle via transfer of oropharyngeal fluid. mSphere. 2018;**3**:1-12. DOI: 10.1128/mSphere.00365-18

[31] Anonymous [Internet]. 2019. Available from: https://pib.gov.in/PressReleaseIframePage.aspx?PRID=1584719 [Accessed: 02 March 2020]

[32] Anonymous [Internet]. 2020. Available from: http://dadf.gov.in/sites/default/filess/Operational%20 Guidelines.pdf [Accessed: 02 March 2020]

[33] Verma AK, Tripathi AK. Investigation of FMD outbreak in district Bulandshahr, Uttar Pradesh, India. International Journal of Livestock Research. 2017;7:113-118. DOI: 10.5455/ijlr.20170124043520

[34] Railey AF, Lankester F, Lembo T, Reeve R, Shirima G, Marsh TL. Enhancing livestock vaccination decision-making through rapid diagnostic testing. World Development Perspectives. 2019;**16**:100144. DOI: 10.1016/j.wdp.2019.100144

[35] Kakker NK, Sharma R. Foot-and-mouth disease outbreaks after the launch of FMD control programme in Haryana. Haryana Veterinarian. 2012;**1**:132-147

[36] Ranaweera LT, Wijesundara WWMUK, Jayarathne HSM, Knowles NJ, Wadsworth J, Gray A, et al. Transboundry movements of foot and mouth disease from India to Sri Lanka: A common pattern is shared by serotypes O and C. PLoS One. 2019;**14**:1-14. DOI: 10.1371/journal.pone.0227126

[37] Singh I, Deb R, Kumar S, Singh R, Andonnisamy J, Smita S, et al. Deciphering foot and mouth disease virus-host tropism. Journal of Biomolecular Structure & Dynamics. 2019;**37**:4779-4789. DOI: 10.1080/07391102.2019.1567386

[38] Das SK, Das P. Indigenous technical knowledge in agriculture. Handbook of Agriculture. New Delhi: Indian Council of Agricultural Research, Directorate of Information and Publications of Agriculture; 2006. pp. 1162-1182